Limits to Terrestrial Extraction

This volume focuses on the social, cultural, and ecological consequences of a political economy of energy.

A political economy of energy holds that an enduring hallmark of the current context is a reorganization of human society toward energy extraction and production. *Limits to Terrestrial Extraction* looks at the construction of society itself as an energy-harvesting "megamachine," the ecomodernist project of the latter half of the twentieth century and its disastrous environmental record, and mining Near Earth Objects to extract extraterrestrial resources. Each chapter explores a limit to terrestrial extraction – spatially, economically, or socially – finding that business as usual cannot yield a different world. The authors eschew easy answers of natural resource management or discourses of wise use, instead offering critiques of market society and its constitutive drive to produce and waste energy. Overall, this volume establishes the existential stakes and scope of change that will be required to build a better world.

This book will be of great interest to students and scholars of environmental political theory, as well as social scientists and humanities scholars who study the intersection of energy and society.

Robert E. Kirsch is an Assistant Professor in the Faculty of Leadership and Integrative Studies at Arizona State University, USA.

Routledge Focus on Energy Studies

Limits to Terrestrial Extraction
Edited by Robert E. Kirsch

For more information about this series, please visit: https://www.routledge.com/Routledge-Focus-on-Energy-Studies/book-series/RFENS

Limits to Terrestrial Extraction

Edited by
Robert E. Kirsch

LONDON AND NEW YORK

First published 2020
by Routledge
2 Park Square, Milton Park, Abingdon, Oxon OX14 4RN

and by Routledge
52 Vanderbilt Avenue, New York, NY 10017

Routledge is an imprint of the Taylor & Francis Group, an informa business

© 2020 selection and editorial matter, Robert E. Kirsch; individual chapters, the contributors

The right of Robert E. Kirsch to be identified as the author of the editorial material, and of the authors for their individual chapters, has been asserted in accordance with sections 77 and 78 of the Copyright, Designs and Patents Act 1988.

All rights reserved. No part of this book may be reprinted or reproduced or utilised in any form or by any electronic, mechanical, or other means, now known or hereafter invented, including photocopying and recording, or in any information storage or retrieval system, without permission in writing from the publishers.

Trademark notice: Product or corporate names may be trademarks or registered trademarks, and are used only for identification and explanation without intent to infringe.

British Library Cataloguing-in-Publication Data
A catalogue record for this book is available from the British Library

Library of Congress Cataloging-in-Publication Data
A catalog record has been requested for this book

ISBN: 978-0-367-86336-4 (hbk)
ISBN: 978-1-003-01848-3 (ebk)

Typeset in Times New Roman
by codeMantra

Contents

	List of contributors	*vi*
1	**Introduction** ROBERT E. KIRSCH	1
2	**Mumford and Bataille: toward a political economy of energy consumption** ROBERT E. KIRSCH	5
3	**Climate change and decarbonization: the politics of delusion, delay, and destruction in ecopragmatic energy extractivism** TIMOTHY W. LUKE	25
4	**Star power: outer space mining and the metabolic rift** EMILY RAY AND SEAN PARSON	54
5	**Conclusion** ROBERT E. KIRSCH AND EMILY RAY	74
	Index	*81*

Contributors

Robert E. Kirsch is an Assistant Professor in the Faculty of Leadership and Integrative Studies at Arizona State University. His research focuses on questions of leadership, citizenship, and authority as well as heterodox economic theory. His next book, *Modern Monetary Theory and Marxism*, is under contract at Routledge for 2020. He can be reached at Robert.Kirsch@asu.edu.

Timothy W. Luke is a University Distinguished Professor in the Department of Political Science and the Government and International Affairs Program in the School of Public and International Affairs at Virginia Polytechnic Institute and State University, Blacksburg, Virginia.

Sean Parson is an Associate Professor in the Departments of Politics and International Affairs, and the MA in Sustainable Communities at Northern Arizona University. He is the author of *Cooking up a Revolution: Food Not Bombs, Homes Not Jails, and Resistance to Gentrification* (Manchester University Press 2018) and has published on topics ranging from Science Fiction and Superheroes to Environmental Political Theory. When he has time, he walks his four-legged best friend, Diego, through the forested mountains of Northern Arizona.

Emily Ray is an Assistant Professor of Political Science at Sonoma State University in Northern California. Her research centers on environmental political theory, with published works on space exploration and climate change, science fiction and political ideologies, and land-use conflicts in North America. Emily can be reached at Emily.Ray@sonoma.edu.

1 Introduction

Robert E. Kirsch

At the 2019 Western Political Science Association annual conference, a group of environmental political theory (EPT) scholars presented work that theorized the politics of waste, energy policy, and energy production in a regime of ongoing climate change. The present chapters that emerged out of the papers on that panel engaged with some fundamental questions about social organization in a political economy of extraction. The paths to address these questions vary: Timothy W. Luke critiques the hollow promises of ecomodernism and ecopragmatism with their unwillingness or inability to challenge the system of production that consumes so much energy. These frameworks promised a brighter future of limitless energy abundance just around the corner, but instead delivered 40 years of devastating environmental impacts; Robert E. Kirsch brings in theorists not often thought of as political economists who contribute to a discussion of the need to profitlessly consume energy. In doing so, he makes the case for why capitalism is ill-suited for such profitless consumption, turning sites of extraction into sites of sacrifice; and Emily Ray and Sean Parson consider a future of asteroid and other Near Earth Object mining, and how such an extraterrestrial push for energy extraction is a continuation of the second contradiction of capitalism by searching for new markets out of Earth's atmosphere. These far-ranging questions from an EPT perspective contribute to a growing body of work in political theory on economies of energy and energy extraction.

These analyses contribute to a growing body of literature in EPT and in political theory generally, of a political economy of energy and/or extraction. While it seems obvious in the current context, the concept of stored energy from material sources as a quantifiable contributor to production that is separate and distinct from human exertion is relatively recent.[1] This conception of energy as a possible external additive to facilitate labor has indeed changed the way that humans perceived themselves as energy-requiring bodies.[2] Far from simply offering policy prescriptions, or best-use practices for already-existing regimes, the critical perspective of many in the EPT community provides for radical analyses of the existential challenges of climate change as they unfold. The contributors to this volume add to this growing discussion and intersect with some of the following urgent themes.

Earthly gods

Luke takes up Stewart Brand's version of ecomodernism with its concurrent politics of ecopragmatism which views the world as a modernist culture that must be contained, disciplined, and bounded. He notes Brand's commandment that "We are as gods, and have to get GOOD at it." Luke critiques this somewhat perverse notion of a stewardship which insists that humans have become the unwitting guardians and, indeed, makers of the planet, arguing that lurking behind them is a regime of control with a ruling class of wise administrators, whose plucky innovation and greening ethos will accept the difficult mantle of remaking the face of the Earth. This will necessitate a wildly disruptive plan of cordoning off large sections of the planet for conservation with little regard to people who live in those places, but that is the price to pay for a "good" global Anthropocene. Ray and Parson discuss a slightly younger set of earthly gods from Silicon Valley. In their chapter, the Silicon Valley innovators are anxious to break free from the terrestrial limits to their ambitions, to set up space colonies, mine asteroids, and begin the new age of interplanetary travel. Ray and Parson offer material reminders that asteroid mining is merely a desire to find new areas of extraction as a way to forego dealing with the environmental disaster of energy extraction on Earth. Their exospheric analysis of Near Earth Objects is grounded in the political contestations of borders, private property, and thus argues that off-Earth industry can only offer a fleeting freedom from the extraction of energy on Earth, and they are dubious that Elon Musk and Jeff Bezos are able to offer anything more concrete. Kirsch uses Lewis Mumford to reach back into the anthropological record to note that societies are often built around producing and storing a surplus of energy; the question is who gets to consume it. Those laying claim to the surplus consume it for their own ends, and often disastrously, and Mumford asks: "As for the great Egyptian pyramids, what are they but the precise static equivalents of our own space rockets? Both devices are for securing, at an extravagant cost, a passage to Heaven for the favored few." Social organization provides the means for the earthly gods of

Brand, Musk, and Bezos to have a surplus of energy to consume as they build monuments to themselves. There is little reason to believe that this arrangement will result in anybody being good at using their godly powers for anything else.

Sites of sovereignty

Each of these contributions makes the case that there is more to extraction than the mere removal of fuel from the ground (or asteroid) for conversion into energy. Kirsch argues that Georges Bataille's theory of a "general economy" shows that however long energy can be converted and stored, because it is itself based on the sun's waste, that energy must eventually stay wasted. Sites of sovereignty are determined by who gets to control the ultimate wasteful consumption of energy, which Kirsch notes gives lie to the notion of productive consumption or constant profitable reinvestment, even in terms of energy production. Ray and Parson take that insight to its terrestrial limit and point out not only the difficulties of private property relations and treaty statuses in extraterrestrial extraction, but also that the carving up of space itself is a Schmittian boundary-drawing. Rather than the frontier that belongs to no one, space colonization is acts of sovereign appropriation. Luke's critique of the ecopragmatists operates on a similar notion of sovereignty when discussing stewardship. They believe that they are the right people to govern the new planetary arrangement, and their vision of stewardship draws new lines of livable spaces and appropriate technologies and ways of life. The stewards take Earth's excess energy and put it to use to sustain a certain kind of lifestyle. While it may be tempting think that once humans are free of the planetary boundaries or have enough energy for a certain kind of life that these sovereign orderings of disciplined subjects will evaporate, but as each author notes, society itself is organized for unrelenting extraction.

The consequences of normal politics

One of the central arguments of Luke's chapter is that while the ecomodernists have been advancing an ecopragmatist agenda, uninterrupted, for almost four decades, not only is there little to show for it, but environmental degradation continues unabated. This is why he suggests that current suites of environmental policy are concerned with delusion and delay, with their attendant destruction being justification for further rounds of ecopragmatism. Ray and Parson argue that the piecemeal ownership of Near Earth Objects by sovereign states will open up new markets in a neoliberal space race; the frontier discourse of outer space is simply the managed absence of the state that makes markets work. Energy problems on Earth will not be solved by finding new sources of energy in space. For Ray and Parson, the arrow goes the other way – the problems of energy production on Earth will merely be exported to Near Earth Objects in space. Kirsch finds that market societies, because they are unable to waste profitlessly, will insist on finding new areas of investment and warns that, historically, this means disastrous insistences on building new markets – through warfare or even through manufacturing destructive sites of energy production. Luke's argument that normal politics does not seek to change the relations of production but rather only change their focus, intensity, or implementation can only delay ecological collapse, not reverse it.[3]

The chapters in this collection ultimately argue in their own way that the current mode of environmental policy cannot lead to a sustainable future unless social relations and the relations of production are fundamentally changed. This is an enormous challenge, and on an increasingly vanishing timeline. Yet the critical engagement of political economies of extraction seeks to accomplish just such a goal.

Notes

1 Cara New Daggett, *The Birth of Energy: Fossil Fuels, Thermodynamics, and the Politics of Work* (Durham, NC: Duke University Press, 2019).
2 Anson Rabinbach, *The Human Motor* (Berkeley: University of California Press, 1992).
3 Timothy W. Luke, *Ecocritique: Contesting the Politics of Nature, Economy, and Culture* (Minneapolis: University of Minnesota Press, 1997).

2 Mumford and Bataille
Toward a political economy of energy consumption

Robert E. Kirsch

Introduction

While Lewis Mumford and Georges Bataille were writing contemporaneously in the mid-twentieth century, it would seem like quite a stretch to suggest that their bodies of literature had an obvious affinity to each other. Indeed, there is very little in the academic literature that puts these thinkers in conversation. While this lack of engagement might suggest that there is nothing to be gained by doing so, this chapter argues that if these two thinkers are thought in explicit terms as theorists of energy, a comparison, and indeed synthesis, becomes a visible and fruitful undertaking. This chapter attempts just such a synthesis and begins with a brief exposition of each thinker as a theorist not only of technology but also of energy. Specifically, it takes up Mumford's conception of technics as an organizing force of society itself for the capture and storage of energy that he called the "megamachine." It also takes up Bataille's theory of a "general economy" in which a given society is dependent on sumptuously consuming excess energy after social reproduction and growth, if it is necessary or possible.

These two strands will be synthesized in a loose contribution to a philosophy of energy and will critique the recent push to open new avenues of terrestrial, and increasingly extraterrestrial, extraction. Most of the excitement around frontiers of extraction is hyped up by Silicon Valley, but Mumford warns that the earthly gods that pilot the megamachine are content to build pyramids to their posterity (or in the current context, rocket ships as Mumford points out), and Bataille is wary of the capitalists that insist on finding "productive" outlets for their surpluses, because the quest for infinite productivity often ends in disaster. The chapter ends with an appeal to a view of energy that should give critical pause to consider whether organizing the megamachine on a global level to find the new frontier of energy production. Rather, it shows the limits of the rationality of terrestrial extraction. By exploring these two thinkers in terms of energy, life-centered economies, sovereignty, and wasteful excess, there is an opportunity to reconceptualize systems of energy rather than the narrow view of extraction-fueled growth for the sake of growth.

Mumford, the myth of *homo faber*, and the megamachine

While Mumford wrote voluminously, and as a generalist, on a number of topics, much of that work has a touchstone in energy.[1] Also as a generalist, he centered his concern around energy not simply on a given deployment energy to run technologies but on energy's relationship to society itself. Part of this task is clearing away the brush of anachronistic Victorian assumptions regarding the anthropological history of humanity, what he refers to as the myth of *homo faber*; the tool user of productive humanity.[2] Critiquing the essentialization of Victorian bourgeois norms as eternal societal was long part of Mumford's intellectual concern, from when he was a student of Thorstein Veblen at the New School for Social Research, and an editor at *The Dial* with Veblen as well.[3] Mumford carried forth and expanded the way in which Veblen critiqued Victorian norms of leisure as an invidious mode of class distinction, recasting the very concept of leisure into a class marker of wealth, nobility, and virtue. Ironically, the leisure class showed a disdain for the labor and wealth became a way to broadcast their virtuous abstention of it as a class marker.[4] The full weight of this insight of leisure being used to flaunt abstention from labor will come to bear below when leisure as a profitless consumption for self-pleasure has effectively been closed off by the need to find "productive" outlets.

Veblen's class distinction of who must work and who has the pecuniary to opt out of that labor is a distinction Mumford seized upon. Mumford expands the scope of the analysis from Veblen's idle rich to the organization of energy. That is, who labors to produce the energy needed for social reproduction, and who is able to appropriate that surplus to their own ends? The question is answered by a corollary of social organization to get those who must work to do so pliantly and build institutions that legitimize this arrangement. Mumford links the conduit of energy into those who have to work and those who do not to an enduring mythos of what humanity is at its core of humans as energy producers.[5] He thus links a theory of society as conduits of energy deployment and the myth of *homo faber* to build his theory and critique of the megamachine. This chapter will not be an attempt to analyze the entirety of Mumford's voluminous life's work, but rather use his writings on technics as a mode of social development and how technics shape or fail to shape a social analysis of technology, which is appropriate given that "technology" is often used to mean technics, tools, certain artifacts, and items that carry an electric charge.[6] To drive down to the way Mumford deployed the term "technology" requires a reorientation in the level of analysis. While the dizzying complexity of the current context often prompts scholars of science and technology in society to focus on artifacts (such as mobile phones, social media portals, appliances, etc.) to give some empirical clarity, it is also important to take on the notion of technology itself and how it, as a way of knowing, shapes the everyday lives of people.[7] Technology is etymologically rooted in the Greek word *technē*, which is a way of knowing by doing, and cultivating a habit of craft mastery as a conduit of personal development.[8] For Aristotle, *technē* was a force for moral development and part of the salutary ethical life of citizenship.[9] Studying technology as a social phenomenon then is how knowledge by doing shapes collective development: a matter-of-fact analysis of production and their concurrent modes of social organization.[10] In a way then, technology is the means of production; how they are organized and what is produced.[11] Studying technology is thus not a discrete section of social forces among others (in this regard, asking "what is technology doing to us?" is an incoherent question), but represents looking at the processes of production, the habits of thought and cultural evolution they engender, and how these relations can be changed to produce other habits of mind or new systems of production. By the same token, energy is the vector by which that production is accomplished. Technology is not a particular set of material relations but is always already intertwined in their social milieus as ways of knowing. Technology does not move about independently, at varying speeds, or intervene in otherwise normal social relations.[12]

This is why Mumford's study of technology develops a general-level theory of technics. His theory of technics is centered around how "*technical* change – or changes in the forms of energy utilization, raw materials consumption, labor and skills, and the technical organization of production – can effectuate important changes in economic and social relations."[13] The study of technology would thus include studying these energy and materials and how the means of production assemble the material, using energy, into the ends of life.[14] Mumford has these relations in mind and their general relation to energy and its use in these ways when he describes the essential processes of life as

> conversion, production, consumption, and creation. In the first two steps energy is seized and prepared for the sustenance of life. In the third stage, life is supported and renewed in order that it may wind itself up... on the higher levels of thought and culture... Normal human societies exhibit all four stages of the economic processes: but their absolute quantities and proportions vary with the social milieus.[15]

The technological apparatus built up for producing not only the goods of life, but indeed for Mumford, life itself, and the energy captured, converted, and consumed to reproduce and expand life (through art and culture), are questions that are answerable through technics.

Mumford uses technics as the level of analysis to distinguish technology from tools and thus ultimately the mind maker of *homo sapien* from the tool user of *homo faber*. Tools are indeed the means by which technics are materially accomplished, but it is technics that guide human development and the tools for it, and it is a mistake to assume tool usage itself is the final cause of human activity. In Mumford's telling, for most of human history, tools were the main way that humans converted their own energy onto the environment around them to furnish their material needs and cultural expressions.[16] Technics, on the other hand, "...denotes the use of machines and the applied techniques of mechanical organization that integrates and monitors the operation of automatically acting machines... as [machines] become more complex through the addition of nonhuman energy sources and vastly more complicated mechanisms..."[17] Technics as a social level of analysis look at the complexity of production, which provides an opportunity to consider the artifacts of technology rather than tools. Whereas the overwhelming complexity in the productive apparatus makes it difficult to describe the means of production generally, focusing in on particular artifacts becomes a way to, if nothing else, gain a conceptual toehold upon which to discuss broader social implications of technology.[18] However, focusing on artifacts, or assuming that technology is only materially present in things that carry an electrical current masks the complexity of the technics that allow these artifacts to come into being in the first place. Mumford is thus shifting the focus to look at artifacts of technology at the level of technics – of social organization itself – to trace the flows of energy which produces artifacts and other ends of life. His theory of technics is a systems-level analysis of the social conduits of energy.[19]

Mumford's technics can be theorized as a matter of energy production, distribution, and consumption inside a complex of machinery that lays the terrain of social possibilities for the development (or stultification) of human culture. Machinery plays a key role in freeing up energy use to be diverted away from simple reproduction into building more lasting cultural edifices. In Mumford's view, the goal of technics should be to make a culture worthy of humanity's aspirations, and energy should be directed for those ends.[20] Indeed, culture is not possible without the ability to contain and store energy, and here Mumford is worth quoting at length:

> The prime fact of all economic activity, from that of the lower organisms up to the most advanced human cultures, is the conversion of the sun's energies: this transformation depends upon the heat-conserving properties of the atmosphere, upon the geological processes of uplift and erosion and soil-building, upon the conditions of climate and local topography, and – most important of all – upon the green leaf reaction in growing plants. This seizure of energy is the original source of all our gains: on a purely energetic interpretation of the process, all that happens after this is a dissipation of energy – a dissipation that may be retarded, that may be dammed up, that may be temporarily diverted by human ingenuity, but in the long run cannot be averted. All the permanent monuments of human culture are attempts, by using more attenuated physical means of preserving and transmitting this energy, to avert the hour of ultimate extinction.[21]

This rather dour contextualization about the final ends of culture nevertheless establishes the necessity for the centrality of systems of energy in social organization. If all energy is in the final analysis dissipated, then the best technics require a social apparatus so more energy can be devoted to building culture. Tools allow humans to expend less of their own energy,[22] and early industrial production used machinery powered by water, steam, or fossil fuels as a way to multiply and duplicate the function of tools in production.[23] As Luke reminds above, Mumford's theory of technics goes beyond simple tool use to see the technological apparatus of machinery and its social and economic consequences. While the impulse toward looking at discrete artifacts may make analyzing machinery a matter of cataloging mechanical objects, for Mumford, the technic of social organization that goes beyond mere tool use to direct and store energy is machinery; machines are structures of social regimentation.[24] Mumford further notes that machine rationality often leads to a view of humanity that is oriented to the machine, and so read that orientation into human nature; thus, *homo faber*, the "tool user."[25] Machine rationality, in other words, has supplanted *homo sapien*, the "mind maker" with a vision of what sets humans apart from other species as its ability to use tools.[26] There is an irony here of how technics supplanted mere tools, but how the resulting technological organization has been deployed to reduce humanity to mere users of tools – or perhaps become tools themselves. Mumford marshals a great deal of anthropological evidence that tool usage was a way to reproduce humanity socially and allow the techniques for artwork to be developed; but the real capacity of humans to think abstractly, produce art, and develop cultural structures by use of those tools is what sets humanity apart.[27] Technics and technology are therefore not merely multiplication of effort by tools, but rather what guides the development and deployment of tools themselves; machinery is not simply additive or multiplicative applications of tools but a matter of social evolution of technics to harness energy to transform matter.

Mumford is quick to point out, however, that by machinery he does not only mean metallurgy, or artifacts that carry an electrical current or run on internal combustion. Rather, he speaks of machinery as regimentation, a social organization that directs the energies of human labor, not to the enhancement of everyday life, but to predictably quantifiable outputs, all through the application of human energy admixed with the social machinery that stores energy in some form:

> The machine I refer to was never discovered in any archaeological diggings for a simple reason: it was composed almost entirely of human parts... Let us call this archetypal collective machine – the human model for all specialized machines, the *Megamachine*.[28]

The reductive myth that humans are mere tool users or tool users of *homo faber* is part and parcel of the sustainability of the megamachine with its "monotonous repetitive tasks" that delivered quite a few communal benefits. For instance, the total social mobilization and organization into the megamachine regulated flood control and agricultural production, guided urban planning and design, as well as a securitized existence owing to the military machine.[29] The discipline of the machine, however, takes an exacting price on the humans directing their energies into its production apparatus, with the megamachine caring only for "speed, uniformity, standardization, quantification."[30]

While such regimentation may seem leveling, Mumford notes, again echoing Veblen's "leisure class," that the ruling classes "claimed immunity from manual labor, were not subject to this discipline," but also, as will be important when discussing Bataille below, "...their disordered fantasies too often found an outlet into reality through insensate acts of destruction and extermination.[31]" This is good evidence for treating the megamachine as a class analysis, and just as important for analyzing who is asked to take part, and who is able to opt out of the labor of sustaining the megamachine. Mumford suggests that the ruling class who can opt out of the labor of the megamachine will often guide it to spectacularly ugly ends. But for now, it will be enough to point out that the ability to abstain from labor and to consume unproductively is the realm of sovereignty, which Mumford points out as well, and Bataille will take that ball and run with it below.

Megamachine culture produces subjects who see themselves as humans fit to participate in the megamachine. This circular rationale obscures the teleological ends of humans of "autonomy, self-direction, and self-fulfillment," and presents a one-sided picture of the technical dictates of the megamachine that frustrates the other areas of human development.[32] For Mumford, the hallmark of modern economic thought is an "unconditional commitment to the megamachine... as the main purpose of human existence."[33] Whereas the fruits of the increased efficiency of social organization could be used for human flourishing and cultural development, Mumford shows instead that the directed energy flows of the megamachine serve only to enlarge itself and the ruling class. In an era where energy extraction may not be limited to *terra firma*, Mumford's seemingly rhetorical question is suddenly less so: "As for the great Egyptian pyramids, what are they but the precise static equivalents of our own space rockets? Both devices are for securing, at an extravagant cost, a passage to Heaven for the favored few."[34] Mumford's ultimately class-based argument about social organization's totality in the megamachine gears social development for the capture and use of excess energy becomes the end of life itself. The question that remains to be answered is what to do with the class distinctions and whether there are different arrangements or aims of the megamachine. Still, Mumford's analysis bears out in the current context, where the ends of life are not a matter of personal or cultural fulfillment, but rather geared toward how to best serve the megamachine – with the ultimate goal being able to ascend to the class that can abstain from labor.

Bataille, solar excess, and sovereign consumption

While the convergences between the two will be more fully explored in the next section, Georges Bataille similarly argued that all human activity, economic or otherwise, is ultimately derived from the sun. His notion of a general economy is thus also one of energy, and one of energy consumption and dissipation, but Bataille more explicitly emphasizes that all of it comes from solar waste.[35] He ties the material conditions of life to the source of energy that makes it possible, by suggesting that the general economy is the capture and storage of a tiny portion of the excess energy from the sun, and temporarily converting it into a material surplus for societal growth.[36] However, once the limit to economic growth is reached, in the long run, "... energy finally can only be wasted" making the conversion of energy temporary or deferred to some point.[37] Solar excess is at the end responsible for the fruitfulness of the land, the ability of labor to manipulate those raw materials, and for a surplus to be accumulated of which capital might lay claim through machinery. Even then, humans are only able to capture, let alone store, a very small portion of that energy to produce their own energy surplus. The rest must be dissipated without profit – without spurring accumulation or further production.[38] An important question then becomes how and who gets to profitlessly consume the surplus energy.

The useless dissipation of that excess solar energy is not a mere physics metaphor about the law of the conservation of energy. Bataille draws some conclusions from this principle of solar energy providing the basis for teeming life. A fundamental one is that because the energy derived from the sun is itself solar waste, even if captured to some degree for some length of time, that energy must remain wasted; its storage is a deferral. It is also helpful to highlight the lexical changes this insight spurs. For Bataille, wealth is the excess solar energy that has been successfully converted or stored to yield a material surplus for system growth.[39] Wealth is thus not productive (or emblematic of productive prowess) but a temporary store of energy to be reinvested some number of times, but will ultimately, and by necessity, be expended without use. Wealth, in other words, is not productive, but consumptive. The benefit of redefining the terrain of the concept of wealth is that it moves away from a kind of mysticism surrounding accumulation of wealth, as if wealth were something that comes from nothing and enlarges itself, and overturns "... the ethics that grounds [it]."[40] Try as the capitalist ethic might to have it otherwise, eventually all that accumulation dissipates.

Reconceiving wealth as a store of energy to be consumed and exuded without profit or reinvestment also denies the traditional economic discourses of factors of production. Instead of assessing which factors are more wealth producing, it is more appropriate to reframe the discussion in terms of outlets for consumption. The ethical dictates of productive consumption, such as the Protestant Ethic, cannot abide the inescapable necessity of wasting excess energy without profit. To be clear, Bataille insists that the energy will be wasted in any eventuality, but not all squandering is the same. Under different modes of production, different kinds of squanderings can be imagined. For Bataille, it is market society that poses the greatest threat. He argues that under the insistence of productive consumption or profitable reinvestment, this excess will be wasted in a damaging way, including the distinct possibility of the annihilation of the human species.[41] In the current context, the inexorable drive to find profitable reinvestment fuels speculative crises,[42] environmental degradation,[43] and massive inequality.[44]

Converting the surplus solar energy into material surplus is necessary for growth and profit, but since it is based on already wasted solar energy, at the same time it establishes a limit to profit in the same way that Mumford offers the inevitability of energy's dissipation. Profits are ephemeral inasmuch as they are unable to be stored or deferred indefinitely, and at some point, the energy must be given up without profit by necessity.[45] This also highlights a brewing problem in political society – if a regime of capital accumulation understands giving up profit to be anathema, what are the consequences of a social lack of avenues for the necessary consumption of excess energy? Bataille discusses religion, luxury, and conquest as outlets, but in an ethos of accumulation for accumulation's sake, the accumulated energy that must be unproductively consumed lacks these outlets. The energy must still be squandered, and may come out in damaging ways under a performative aegis of productive use of excess energy.[46] Thus, when considering a political economy of extraction, sites of energy production can be recast as sites of squandering excess energy under the guise of productive reinvestment, even if they cannot recognize themselves as such.[47] Doing so makes sense of the environmental, social, and human damage of extractive sites, reconceiving them not as places to profitably attain more energy, but as sites where surplus energy is squandered at great cost.

Growth, related to wealth, also undergoes a substantive lexical change in the general economy. Growth is a social category that represents a surplus that has been consumed beyond a given society's needs but is not yet wasted, and as such may be profitably reinvested given certain social conditions.[48] That is to say, the energy needs of society vary based on development and historical era. The difference between social reproduction and surplus deployment, for Mumford, is the extent to which human culture may be expanded. Yet growth also has a limit, to use Marxist terms, based on the development of the constant and variable capital in a society. Either the development of constant capital eventually reaches a point where it is unable to grow anymore and must create new spaces to absorb it, which is the crux of Bataille's critique of the Marshall Plan, or a society is not yet able to capture enough solar excess to expand the constant capital enough while maxing out labor-power, which is the crux of Bataille's critique of Stalinism. This new register of growth taken in concert with wealth sheds an important insight on a crisis of overproduction. Overproduction represents material surplus beyond the capacity for growth/economic development to absorb it. If this surplus cannot be absorbed or reinvested, again, it must be squandered. According to Bataille, the balance of material after the limits to growth have been reached would necessarily need to be squandered, but in a capitalist society, there are no clear outlets for the exudation. Bataille notes that the squandering must happen, with sometimes disastrous consequences, such as war – and further notes that development plans, such as the Marshall Plan, provide a warlike outlet for excess without battle.[49] It seems that this quasi-militarization and deployment of productive forces to new areas of the world that had been damaged in world wars, and not necessarily a cornucopian abundance, is the rationale behind Bataille's statement: "...what matters *primarily* is no longer to develop the productive forces but to spend their products sumptuously."[50] It is clear that for Bataille, a society that reordered itself on consumption of excess, rather than insisting on production and reinvestment to battle scarcity, would be a society that could squander that excess on luxury. Bataille is not asking for wastefulness as response to a crisis of overproduction, but rather wastefulness as sumptuously consuming the excess energy that has been stored through production in the present, without an eye on profitable reinvestment in the future.

This reorientation of wealth and growth as vectors of one's ability to consume rather than one's propensity to produce shows how destructive sumptuous consumption in the general economy can be. Because it is the result of a political society unable to see its need to nonproductively consume its excesses, lacking the necessary social outlets, it does so in destructive ways.[51] Bataille points out that warfare is often an outlet when profitless consumption is hoarded by the few:

> Under present conditions, everything conspires to obscure the basic movement that tends to restore wealth to its function, to gift-giving, to squandering without reciprocation. On the one hand, mechanized warfare, producing its ravages, characterizes this movement as something alien, hostile to human will.[52]

Because the squandering is not done sumptuously, by and for the people producing the surplus, the consumption of the surplus is alienating, and often something dangerous and destructive.

This factor of alienation from sumptuous consumption in a political economy of extraction is complementary to the "second contradiction" argument that eco-Marxists have made,[53] Bataille's emphasis on the general economy highlights the necessity of manufacturing that alienation from nature, though not necessarily by highlighting the environmental degradation itself. Rather, he focuses on how productive consumption forecloses avenues of absorbing surplus or uselessly consuming excess by instead insisting on making extraction a "productive" enterprise. This is the danger of an economic ethos that does not allow for useless consumption, Bataille argues, and must be overturned.[54] A factor of alienation in capitalist society then is a lack of outlets for useless consumption of the surplus beyond what is needed for growth. Overcoming that alienation is to restore the ability of those who produce the surplus to also squander it.

The alienation that is manufactured in this framework is not just humanity from nature, where a unity of opposites might point the way for a kind of homecoming. Of course, it is that but it is also the alienation produced in an ethic of accumulation for accumulation's sake is also the alienation of humanity from enjoying what it produces by deferring consumption. Here it is useful to draw upon Marx's theory of alienation, which Bataille explicitly relies on, to highlight how reducing human activity to a constructed set of productive factors alienates humanity from itself. Marx's two-part argument of alienation is not only that humanity is estranged from nature via the labor processes, but also that the productive processes under capital accumulation alienate humanity from its species-being.[55] It is to this second aspect that Bataille complements, by placing the realm of sovereign (or non-alienated) consumption in the ability to uselessly consume with regard only for the present after producing for socially necessary wealth and growth. This is in direct contrast to how capitalist society compels deferring consumption for profitable reinvestment or further production, reducing laborers into things, alienated from the things that they produce. For Bataille, sovereign power is the ability to squander, with no concern for anything but the present.[56] Alienation of humanity from what it produces is deferring this sovereign exudation for a rational production of things from those that produce them, thereby reducing humanity to a thing, since the deferral of both sumptuous consumption and growing the means of production do not allow for consumption of the things being produced.[57] This echoes Mumford's critique of the rationality of the megamachine transforming notions of humanity's higher purpose as merely being participants in the megamachine. In Bataille's reading of Marx, Marx's attempt to strip away all nonmaterial vestiges of the production process was to show how humanity, in a regime of accumulation for accumulation's sake, is in the service of things, rather than things being in the service of a fulfilled humanity.[58] In this regard, when Bataille posits "the fulfillment of things," he means this in a radical way, that strips away the fantasies of neoclassical economics and ethical prompts like utility, and scarcity, to reconceive of consumption as something to be luxuriated in after socially necessary production and growth have been achieved.[59] This points toward a realm of freedom, where humans produce for their needs, and consume what is left over and the demands of sustenance and growth are met. Overcoming alienation is the freedom to dispose of the surplus without regard to deferring to the future.[60] If humanity is not free to do this when necessary, the surplus will still be exuded, but the useless consumption will itself be alienated because of the insistence on productive consumption, resulting in a faux sovereignty for those consuming the excess supposedly productively, and a faux freedom for those who are subject to that (creative?) destruction to partake in the endeavor. Or perhaps the worker will have to take solace in fleeting moments of facsimiles of sovereignty:

> If I consider the real world, the worker's wage enables him to drink a glass of wine: he may do so, as he says, to give him strength, but he really drinks it in the hope of escaping the necessity that is the principle of labor. As I see it, if the worker treats himself to the drink, this is essentially because into the wine he swallows there enters a miraculous element of savor, which is precisely the essence of sovereignty. It's not much, but at least the glass of wine gives him, for a brief moment, the miraculous sensation of having the world at his disposal.[61]

In Bataille's political economy of consumption, labor without reward awaits those who are not part of the ruling class' ability to squander energy, but he adds that even those who are in the ruling class have an ersatz sovereignty because they operate under an ethic of productive reinvestment and not sumptuous enjoyment.[62] Bataille claims ultimately that the capitalist class are confused: they do not know why they waste, and because of that ignorance they do so in a harmful way, but the imperative of an economy of productive consumption renders them "helpless" to act any other way.[63] This reinforces the class element of energy that Mumford lays out, but in even more stark terms; the members of the laboring and ruling classes are, to again use Marxian terms, bearers (*trägers*) of their class positions in energy production and consumption. Bataille mentions war and the manufacture of being perpetually on the brink of war, but this alienated sovereignty of the few also finds expression in the catastrophe of financial "innovation" – it destroys excess, but destroys the everyday lives of those who are not able to participate in this destruction.[64] In other words, the nonproductive consumption must be dressed up to look productive, and the sovereign exudation being undertaken by those who are not doing the exuding find outlets of a facsimile of sovereign expenditure.[65]

Indeed, the destructive, quasi-sovereign exudation afforded to the few is done on the backs of the many. It is here that Bataille finds the grounding for a materialist leftist politics. He argues that working-class politics are, in the final analysis, opposed to capitalism inasmuch labor wants a "... greater share of wealth devoted to nonproductive expenditure."[66] In other words, labor wants to squander the wealth that it produces. The political struggles of labor in a market society can then be recast in terms of fighting over the right to squander surplus energy. Some of the ways in which left politics won its share of wealth to consume are very obvious, such as weekends, but Bataille goes further still and suggests that in this class struggle the capitalist class would rather exude that excess wealth, however catastrophically, than cede it to everybody and give up its class distinction. Going back to the Marshall Plan, Bataille notes that the Marshall Plan was a gift only in the sense that the United States needed to find a place for its surplus wealth, lest "...the world's fever rise."[67] Indeed, the economic interventions and Structural Adjustment Programs of the IMF and World Bank can similarly be seen less as development projects or gifts than as a place to funnel surplus wealth productively.[68]

Without a planned outlet for surplus, the exuded wealth is released destructively.[69] This destructive extraction must be confronted because, as Bataille notes, the exudation happens regardless of intent or desire, and so, "It is only a matter of an acceptable loss, preferable to another that is regarded as unacceptable: a question of *acceptability*, not utility."[70] What remains is to examine a site of production/consumption and assess its acceptability, not its productivity. This is a different way to think about extractive political economies, which focus less on their usefulness or returns, and more about how acceptable they are with the human and social costs they carry.

Synthesis and plenty

The side-by-side analysis of Mumford and Bataille around a political economy of energy presents a strong logic for a systems-level approach to energy and extraction, and, in doing so, challenges the logic of limitless growth and extraction. One could frame this as a mere technical challenge to better harness the enormous amount of wasted energy the sun exudes every day (in fact it is doubly wasted – wasted by the sun as well as wasted by humans' failure to capture it), or are there systemic demands of the quasi-sovereign earthly gods steering the megamachine that make productive energy extraction part of a pressure release valve for overaccumulation? In other words, the question must be qualitative rather than quantitative: "Why exactly would our lives become better should we command an even greater amount of energy?"[71]

Both Mumford and Bataille firmly establish that all energy for all human activity (here, Bataille may be a bit more rhetorically precise with his notion of the general economy) comes from solar excess; waste from solar nuclear fusion. They also both have a view of sovereignty as a force of an earthly god in the form of kings and capitalists. It should be noted that neither uses the term "sovereignty" strictly in the sense that comes from Hobbes's political theory, but rather as an organizational component of energy production and consumption given preferred relations of production. In both Mumford and Bataille, the sovereign is the only entity who can consume without producing. For Mumford, the sovereign is who can build, direct, and wield the megamachine without participating in its rigid monotony and strict division of labor. For Bataille, the sovereign is who can sumptuously consume with regard to future production. Both also point out the fact that energy will be dissipated ultimately, regardless of how it is captured, redirected, or squandered. As a result, both thinkers note that a different social organization is needed in order to use energy in a different way.

The main seeming difference between the two is what to do with the surplus energy that is harvested or captured before its ultimate squandering. Mumford would use it to build a material culture worthy of the best aspirations of life.[72] Bataille would "embrace the burn-off of surplus energy that makes our existence possible."[73] Here, the synthesis of these two theorists converges around the sources of energy but diverges around its ultimate consumption. They have different emphases but are not contradictory. Perhaps most interestingly is this divergence about the ends to which surplus energy may be put. Mumford believes that without the megamachine, or at least with different social machinery that is geared toward fulfilling human life, the advances in technics can lead to a broader human flourishing. He directly links the pyramids of Giza to a Keynesian busywork program, and that rocket ships are the current context's version of pyramids for the leisure class.[74] However much Bataille may agree with the idea of a megamachine being primed for destruction with no outlet valves for the surplus, Bataille did not seem to think that there were any cultural pools to drain this energy into either.[75] The tension here seems to be a philosophy of life itself. Mumford sees the capacity for surplus energy as the *sapien* part of humanity's species, and thus argues for life-affirming ways to best preserve that surplus even if it is ultimately squandered; for Bataille the life-affirming way to manage the surplus is only to squander it, reorienting subjects in the present, and reasserting sovereign supremacy over the objects of life. The goal is not to overcome this difference, but rather to integrate these two strands to think about energy as a system that allows for a more egalitarian culture-building and squandering rather than a machine that allows a few to abstain and squander all of that surplus themselves with potentially catastrophic consequences. Both poles of living in the present via squandering excess as well as diverting energy into lasting cultural development are part of political economy of energy.

The strength of this synthesis is also strengthened by pulling these theorists out of their disciplinary "homes" to put them into conversation. By focusing on Bataille and Mumford as political economists, emphasis is shifted away from Bataille as an ethical theorist and Mumford as an archaeologist/urban designer to mine new insight. While some scholarship on Bataille's theory of energy has focused on the ethics of energy expenditure, that there is a good and bad "duality" to how energy is consumed or wasted, the present analysis takes this consideration out of the ethical realm.[76] Rather, a focus on political economy centers the myth of profitable reinvestment, or productive consumption, and the classes of people who create and consume the energy surplus. Energy consumption is only good or bad to the extent that it is defined in its class roles, and even then, as both Bataille and Mumford theorize, the class positions of a given society are not decisions one makes. At the same time on the social level, taking Bataille's analysis of the Marshall Plan and Stalinism as social iterations of the good and bad duality of energy consumption is similarly misguided. While it may be tempting to argue that the Marshall Plan is an example of good energy consumption,[77] Bataille the political economist shows that the Marshall Plan was no gift to the world. Rather, it was an investment opportunity that emerged only after the warfare megamachine and its systematic destruction which provided the "external" space where capital could be profitably redeployed to redevelop parts of the world. Other "gifts" like financial bailouts also create space for reinvestment but do little to squander excess energy. Indeed, when looking at the Great Financial Crisis of 2008, the ability to sumptuously consume unimaginably large quantities of bailout money was given not to homeowners but to the same predatory bankers that blew the bubble in the first place.[78]

Similarly, the appeal to life itself that Bataille makes can pull Mumford out of being treated only as an urban planner or architect. Mumford's megamachine was not merely a fanciful reconsideration of social organization, but a totality of control that operates with frightening capacity to end human life on the planet.[79] While perhaps Mumford's concern with thermonuclear exchange is no longer the most pressing threat (though still very real), geoengineering and extraterrestrial extraction pose the same threats of a megamachine ramped up with awful destructive force with new "gods of the Anthropocene."[80] Mumford is not handwringing when he decries the automation of knowledge – not because technics of knowledge production are good or bad, but because in the milieu of service to the megamachine, knowledge production and science are construed to perpetuate the megamachine.[81]

Conclusion

This chapter has taken an initial step in interpreting Lewis Mumford and Georges Bataille as political economists of energy; and there is surely more theorizing to do as political theories of energy develop. Still, by insisting on a conversation between these two theorists and analyzing their commonalities and divergences, there is room for a critical analysis of energy extraction that simultaneously challenges the ways in which surplus energy production is channeled, with Mumford asking to focus on more fulfilling life-directed ends.

Meanwhile, Bataille challenges the very notion of putting a surplus to good use and instead theorizes a general economy of energy where consumption is the way to activate sovereign control of objects and provide a real sense of the present by squandering the excess energy, which in capitalist society is reserved for sovereign power. Both of the poles, building culture for a future and being able to sovereignly claim the present for one's own everyday life, are a vital component of this synthesis. A political economy of energy that resists the megamachine and lives in the present is one worth building, and those twin poles are most clearly apparent when engaging Mumford and Bataille together.

Notes

1 Robert-Jan Geerts, Bart Gremmen, Josette Jacobs, and Guido Ruivenkamp, "Towards a Philosophy of Energy," *Scientiae Studia* 12, no. SPE (2014): 105–27, doi:10.1590/S1678-31662014000400006.
2 Lewis Mumford, "Tool Users vs. Homo Sapiens and the Megamachine," in *Philosophy of Technology: The Technological Condition: An Anthology*, ed. Robert C. Scharff and Val Dusek (Malden, MA: Wiley-Blackwell, 2014), 381–88.
3 Donald L. Miller, *Lewis Mumford: A Life* (New York, NY: Grove Press, 2002).
4 Thorstein Veblen, *The Theory of the Leisure Class* (New York, NY: Oxford University Press, 2009).
5 Lewis Mumford, *The Myth of the Machine: Technics and Human Development* (New York, NY: Harcourt, 1967).
6 Timothy W. Luke, *Social Theory and Modernity: Critique, Dissent, and Revolution* (Thousand Oaks, CA: Sage, 1990).
7 Peter-Paul Verbeek, *Moralizing Technology: Understanding and Designing the Morality of Things* (Chicago, IL: University of Chicago Press, 2011).
8 Aristotle, *Nicomachean Ethics*, trans. Terence Irwin (Indianapolis, IN: Hackett Publishing Company, 1999).
9 Tom P. S. Angier, *Techne in Aristotle's Ethics: Crafting the Moral Life*, Continuum Studies in Ancient Philosophy (New York, NY: Continuum International Publishing Group, 2010).
10 Thorstein Veblen, *The Theory of Business Enterprise* (New York, NY: C. Scribner's Sons, 1915).
11 Luke, *Social Theory and Modernity*.
12 While it goes beyond the scope of this paper, it is worth noting that the academic discipline of economics is at least partially responsible for thinking of technology exogenously. In the dominant mode of neoclassical economics, "markets clear," meaning that every seller of commodities has a buyer at a certain price. With this model of equilibrium in mind, financial crises are an impossible outcome in these general equilibrium models (Minsky 2008; Davis, 2009; Henwood, 2003), and technology – or technological innovation – is deployed as exogenous shocks which cause crises in markets so that they have to find a new equilibrium as markets

adjust to the new prices wrought by the new artifacts of technology (Schumpeter, 2008). Indeed, some economists have offered that technological development needs to be its own factor of production (Hippe, 2013). It seems a little absurd to suggest that a phenomenon outside of production should be included as a factor of that production, but the discipline has a rather substantial interest in keeping financial crises exogenously created.

13 Luke, *Social Theory and Modernity*, 22.
14 Melinda E. Cooper, *Life as Surplus: Biotechnology and Capitalism in the Neoliberal Era* (Seattle: University of Washington Press, 2008).
15 Lewis Mumford, *Technics and Civilization* (New York, NY: Harcourt Brace & World, 1962), 375.
16 Mumford, *The Myth of the Machine*.
17 Luke, *Social Theory and Modernity*, 28.
18 Geerts et al., "Towards a Philosophy of Energy."
19 Geerts et al.
20 Geerts et al.
21 Mumford, *Technics and Civilization*, 375.
22 Mumford, "Tool Users vs. Homo Sapiens and the Megamachine."
23 Thorstein Veblen, *Essays in Our Changing Order* (New Brunswick, NJ: Transaction Publishers, 1997).
24 Mumford, *Technics and Civilization*.
25 Mumford, "Tool Users vs. Homo Sapiens and the Megamachine."
26 Mumford, 384.
27 Mumford, *The Myth of the Machine*.
28 Mumford, "Tool Users vs. Homo Sapiens and the Megamachine," 385.
29 Mumford, 385–86.
30 Mumford, *The Myth of the Machine*, 239.
31 Mumford, *Technics and Civilization*, 182.
32 Mumford, "Tool Users vs. Homo Sapiens and the Megamachine."
33 Mumford, 386.
34 Mumford, *Technics and Civilization*, 12.
35 Georges Bataille, *The Accursed Share: An Essay on General Economy, Vol. 1: Consumption*, trans. Robert Hurley (New York, NY: Zone Books, 1991).
36 And the energy from the sun is itself wasted in its process of fusion when hydrogen atoms are compressed into helium, and some of the mass of the hydrogen atoms is converted into light energy. All life is thus due to the sun's nonproductive expenditure of energy during nuclear fusion.
37 Bataille, *The Accursed Share*, 10–11.
38 The miniscule amount of solar energy actually captured is evidence for this, of course, but in the broader sense of energy for life, at some point the energy is expended uselessly in death.
39 Bataille, *The Accursed Share*, 21.
40 Bataille, 25.
41 Geerts et al., "Towards a Philosophy of Energy."
42 Earl Gammon and Duncan Wigan, "Veblen, Bataille and Financial Innovation," *Theory, Culture & Society* 32, no. 4 (July 1, 2015): 105–31, doi:10.1177/0263276414566643.
43 Timothy W. Luke, *Ecocritique: Contesting the Politics of Nature, Economy, and Culture* (Minneapolis: University of Minnesota Press, 1997).

44 Thomas Piketty, *Capital in the Twenty-First Century*, trans. Arthur Goldhammer (Cambridge MA: Belknap Press, 2014).
45 Bataille, *The Accursed Share*, 28–29.
46 Bataille, *The Accursed Share*.
47 Robert Kirsch, "Manufacturing Alienation in the Bakken: Toward a Political Economy of Extraction," *Theory & Event* (forthcoming) (n.d.).
48 Bataille, *The Accursed Share*, 45.
49 Bataille, 187.
50 Bataille, 37.
51 Geerts et al., "Towards a Philosophy of Energy."
52 Bataille, *The Accursed Share*, 38.
53 James O'Connor, "On the Two Contradictions of Capitalism," *Capitalism Nature Socialism* 2, no. 3 (October 1, 1991): 107–9, doi:10.1080/10455759109358463; John Bellamy Foster, Richard York, and Brett Clark, *The Ecological Rift: Capitalism's War on the Earth* (New York, NY: Monthly Review Press, 2011); Sarah M. Surak, "Capitalist Logics, Pollution Management, and the Regulation of Harm: Economic Responses to the Problem of Waste Electronics," *Capitalism Nature Socialism* 27, no. 1 (January 2, 2016): 106–22, doi:10.1080/10455752.2015.1136663.
54 Bataille, *The Accursed Share*, 25.
55 Karl Marx, *Early Writings* (New York, NY: Penguin Classics, 1992), 328.
56 Bataille, *The Accursed Share*, 63.
57 James K. Rowe, "Georges Bataille, Chögyam Trungpa, and Radical Transformation: Theorizing the Political Value of Mindfulness," *The Arrow* 4, no. 2 (May 2, 2017): 47–69.
58 Bataille, *The Accursed Share*, 135.
59 Bataille, 140.
60 Georges Bataille, *The Accursed Share, Vols. 2 and 3: The History of Eroticism and Sovereignty*, trans. Robert Hurley (New York, NY: Zone Books, 1993), 214.
61 Bataille, *The Accursed Share, Vols. 2 and 3*, 199.
62 Bataille, 141.
63 Bataille, *The Accursed Share, Vols. 2 and 3*, 141.
64 Gammon and Wigan, "Veblen, Bataille and Financial Innovation," 120.
65 A worker may have her wine, but a capitalist might blow the top off a mountain, for the same feeling of sovereignty.
66 Bataille, *The Accursed Share*, 154.
67 Bataille, 175.
68 Mike Davis, *Planet of Slums* (New York, NY: Verso, 2007).
69 Geerts et al., "Towards a Philosophy of Energy."
70 Bataille, *The Accursed Share*, 31.
71 Geerts et al., "Towards a Philosophy of Energy," 115.
72 Lewis Mumford, *The Conduct of Life* (New York, NY: Harcourt Brace, 1951).
73 Geerts et al., "Towards a Philosophy of Energy," 115.
74 Mumford, *The Myth of the Machine*.
75 Bataille, *The Accursed Share*. In particular, his chapters on the Marshall Plan and Stalinization.
76 Allan Stoekl, *Bataille's Peak* (Minneapolis: University of Minnesota Press, 2007), 54.

77 Stoekl, 54.
78 Gerald F. Davis, *Managed by the Markets: How Finance Re-Shaped America* (New York, NY: Oxford University Press, 2009).
79 Mumford, *The Myth of the Machine*, 228.
80 Bronislaw Szerszynski, "Gods of the Anthropocene: Geo-Spiritual Formations in the Earth's New Epoch," *Theory, Culture & Society* 34, no. 2–3 (2017): 258, doi:10.1177/0263276417691102.
81 Lewis Mumford, "The Automation of Knowledge," *AV Communication Review* 12, no. 3 (September 1, 1964): 261–76, doi:10.1007/BF02769061.

3 Climate change and decarbonization

The politics of delusion, delay, and destruction in ecopragmatic energy extractivism

Timothy W. Luke

Introduction

This critical analysis of energy extractivism examines the key thought leaders at the forefront of what is identified, in terms of an environmental style of thought and action, as "ecopragmatism" as well as "ecomodernism." With respect to the ecopragmatist school of thought, it focuses on Stewart Brand, an eclectic, if not eccentric, entrepreneurial and intellectual force in the California counterculture since the 1960s. Ecomodernist thinkers, in turn, are represented by individuals tied to the Breakthrough Institute, Ted Nordhaus and Michael Shellenberger, who proclaimed "the death of environmentalism" in 2004 in their efforts to depict "a post-environmental world."

To counter global warming, many focus on how to curtail and/or end the practices of "extractivism," which entangles many economies and societies around the world in the unrelenting extraction of commodified natural resources, like coal, gas, and oil as well as minerals, timber, and foodstocks, from underdeveloped regions to more developed zones for more value-added processing, resale, and use in intricate relations of unequal exchange.[1] Even though ecopragmatism appears to oppose such activities in its climate change resistance, its ecomodernist designs focus instead on changing only the extent, intensity, and type of extractivist practices, which lead to a politics of delusion, delay, and destruction in terms of policy impact and implementation.

This chapter surveys these new schools of thought and links them, first, to the traditional foci of twentieth-century environmentalism, from preserving wilderness to conserving wildlife and its habitats, and, second, to the complex challenges of rapid climate change as they have been studied since the 1980s. Yet one must ask if anything is truly new and distinctive in ecomodernism, given its strong attachment to many high modernist technological, operational, and political ideals from the mid-twentieth century. Indeed, a key ecomodernist aspiration is to create "a good Anthropocene," and those strategies appear only to shift extractivism from fossil fuels in a carbon capitalist economy through decarbonization policies that stress renewable energy supplies. Still, the renewability of those energy sources after "decarbonization" aims at exploiting different mineral resources for a new era of "radiationization." In forsaking fossil energy, ecopragmatism turns to a plutonium, thorium, or uranium economy to make good on ecomodern promises to maintain global capitalism on a more equitable and sustainable basis.

Climate change challenges and environmental politics

The *ultima ratio* of contemporary climate change activism is to develop and deploy strategies for the "decarbonization" of advanced industrial economies and societies to cap global warming at no more than 3.6°F (2°C) above preindustrial global temperatures. Many plans for how to attain this goal have been advanced since the late 1980s. No comprehensive, enforceable, and workable program for launching these comprehensive technology, policy, and culture changes, however, has gained any effective traction for decades.

Former NASA scientist James Hansen issued what many regard as "the first warning to a mass audience about global warming," as reported to "a US congressional hearing he could declare 'with 99% confidence' that a recent sharp rise in temperatures was a result of human activity" during June 1988.[2] As carbon emissions in 1988 rose from 20 billion tons to over 32 billion tons three decades later, in 2018, Hansen ruefully noted, "All we've done is agree there's a problem" to the extent that activists and scientists

> agreed that in 1992 [at the Earth summit in Rio] and re-agreed it again in Paris [at the 2015 climate accord]. We haven't acknowledged what is required to solve it. Promises like Paris don't mean much, it's wishful thinking. It's a hoax that governments have played on us since the 1990s.[3]

On the one hand, layers of unconscious habit, heavy capital investment, and deep dependence make change difficult; but, on the other, "fossil fuel companies such as Exxon and Shell" were conscious enough of the perils of rapid climate change years before Hansen's and other scientists' 1988 congressional testimonies "to support a network of groups that ridiculed the science and funded sympathetic politicians" to favor fossil fuels over alternative energy sources.[4] Indeed, recent studies of the oil majors during the 1970s document how they researched the inevitability of climate change, developed corporate adaptation and mitigation strategies in response, and then suppressed the science (except to plan their own endangered capital investments) to maintain profitability, avoid regulation, and stall litigation.[5]

One tactic for ecological critics today has been to advance "pragmatic approaches" in changing fossil-fueled economies, even though there is little consensus about what is realistic, sensible, or practical when facing these titanic policy changes that could have been made less disruptively 50 years ago. Nonetheless, some practically inclined actors reason that deeply entrenched patterns in capitalist exchange are best met by such pragmatic responses, which would spin up conventional counterforces, like "natural capitalism," "green business," or "corporate sustainability," opposing "business as usual" with "disruptive business" to initiate change.[6]

Hence, networks of self-proclaimed commonsensical thinkers, sometimes tagged as "ecopragmatists," and other times identified as "ecomodernists," emerged during the 1990s and 2000s to popularize putatively hard-nosed and business-like solutions to address rapid climate change and its associated environmental degradation. Like Hansen, they seem to oppose the unrestrained resource-exhaustive mode of extractivist political economy underpinning contemporary global capitalism as it relentlessly exploits more energy, land, materiel, and labor year after year. They also decried how it produces more commodities with greater waste, less equity, and more pollution, but they also believed they could salvage its lifestyles by finding pragmatic paths to truly green modernization. As the Chinese ideogram for "crisis" on the cover for Michael Schellenberger's and Ted Nordhaus's extraordinary 2004 manifesto, "The Death of Environmentalism: Global Warming Politics in a Post-Environmental World," by linking the characters for "danger" and "opportunity," these thinkers see today as one of history's most pivotal moments.[7] How then do ecomodernists frame the dangers, and ecopragmatists leverage opportunities, as climate change continues to worsen, to improve everyday life?

Breaking through: ecomodernism and ecopragmatism

As a manifesto issued first in October 2004 at an October 2004 meeting of the Environmental Grantmakers Association (EGA), one must wonder if "The Death of Environmentalism: Global Warming Politics in a Post-Environmental World" basically has only deflected or diffused fears about today's crisis with their ecomodernist program. Are the leading lights in these two key networks of ecopragmatism and ecomodernism, like Brand, Elkingon, Hawken, or the Lovins, instead trapped within the rhetorical capture regimen of the actually existing corporate environmental policy, philanthropy, and publishing industries rather than developing effective developmental solutions to advance global warming politics in their allegedly post-environmental world?

At their debut at 2004 Fall retreat of the EGA, which is a "hallmark event" for major environmental lobbying networks, the ecomodernist "post-environmentalism" of the Breakthrough Institute appeared pragmatically to match their tactics to the instrumental eco-managerialism of the EGA. That is, they too "organized small and large events across the country to help its members learn, share, network, and collaborate" with firms doing "business as usual" to consider the pragmatics of green modernizationist thinking.[8] Their disruptions were minimal despite their contrary claims.

In contrast, Rachel Carson published her powerful *Silent Spring* expose in the second year of the Kennedy White House. Over the next 15 years, her environmental tract had triggered tremendous cultural, political, and social changes from a new ecology-minded counterculture, the establishment of Earth Day, new activist environmental groups, and a series of environmental protection actions with executive orders, acts of Congress, and judicial rulings during the Johnson, Nixon, and Carter White House years. Fifteen years after announcing the age of "post-environmentalism," it should be well underway; however, the ecomodernists working with the Breakthrough Institute have done little more than enable more delusion, delay, and destruction by packaging their extractivist-friendly ideas as the "breakthrough" against "the old environmentalism" unable to realize the decarbonization of the world economy and society.

At best, however, this ecopragmatic "new environmentalism" has only called attention to its own Anthropocene-branded green politics by encouraging human beings to "love their monsters,"[9] embrace their "politics of possibility,"[10] and lobby for the Trump White House's endorsed and Republican Senate supported Nuclear Energy Leadership Act (NELA) to get two to five advanced Nuclear technology reactors constructed by 2025 as a piece of their "smarter innovation policy." To top off these moments of rhetorical capture by today's actually existing extractivist capitalism, they invited the world to their Ninth Breakthrough Dialogue during June 2019 in Sausalito, California to fete the ecopragmatist thinker, Stewart Brand, who they regard as "the original ecomodernist, on the 10th anniversary of the publication of *Whole Earth Discipline*."[11]

Meanwhile, Michael Shellenberger has launched his pro-nuclear power auxiliary operation, Environmental Progress, which tracks the planning, construction, operation, decommissioning, and productivity of civilian nuclear energy generating stations. In clear opposition to the "old environmentalism" that came into the 1960s with a preservationist agenda that espoused resource conservation, "the politics of possibility" for Shellenberger and Nordhaus today implies constructing entirely new built environments to replace fossil fuels with a nuclearized energy infrastructure. Without much thought being given to closing the still open nuclear fuel cycle in terms of long-term waste storage, short-term reprocessing costs to refuel reactors, chronic challenges in terms of public safety with regard to transporting nuclear materials today much less on a scale two to three times greater to replace fossil fuel electricity generation, and the command/control/communication requirements of a much larger nuclearized political economy, there seems to be a "pragmatic deficit" in ecomodernism.

Admittedly, ecomodernism has its attractions. Despite the recurrent energy crises triggered by OPEC in the 1970s, and years of stagflation during the 1980s in the USA, worldwide levels of growing industrial pollution and production continued rising during the Cold War era, especially as Deng Xiaoping's "Four Modernizations" kick-started China's rapid industrialization after 1978. By 1988, as climate scientists detected unusual patterns of rapid atmospheric warming, very respected scientific experts, like James Hansen and Steve Schneider, stressed the urgent need for decarbonizing human energy use. This change was imperative to avoid, or at least adapt to, major climate changes around the planet, which were being connected to "the end of Nature" by Bill McKibben, who painted a painful picture of a coming apocalypse that pragmatic "climate change policy" sought to manage.

For the most part, many green thinkers since the 1960s from Rachel Carson, William Ophuls, Barry Commoner, and Edward Abbey to Paul Ehrlich, Murray Bookchin, Donella Meadows, and Garrett Hardin were apocalyptic brand builders. Often arguing that the ultimate catastrophe of a dead Nature was quite near and still closing fast, Shellenberger and Nordhaus pivoted their rhetoric of catastrophe against them to declare instead "the death of environmentalism."[12] Naomi Klein's dreary works about global markets captured by corporate logo consciousness, states baffled by the imperatives of disaster capitalism, and new scientific data that "changes everything" deepened the ecomodernists' caricatures of many green thinkers.[13]

30 *Timothy W. Luke*

By endlessly wallowing in the doom and gloom of inevitable climate catastrophe, Klein perfectly exemplifies the leftish chiliastic climatology that pushed ecopragmatists to become the default ethico-political "center" of today's ecomodernizing designs for "saving the planet" without worrying about the high old school "rightish" hard green conservationist criticism of environmentalist from the 1990s.[14]

The ultimate ecopragmatist

To grasp the scope and substance of the ecopragmatist turn, one can turn to earlier works and proclamations, like the ecopragmatism in Stewart Brand's *Whole Earth Discipline*.[15] Most importantly, these notions underscore, once again, how completely "sustainable development," which at one time was panned as a strange counterculture delusion, has stayed alive and well in the USA as "the triple bottomline" for much of corporate America during and after the Dot-Com Bust and Great Recession.[16] The endless resourcification of the Earth and its inhabitants that lies at the core of industrial – capitalist, fascist, or state socialist – extractivism crystalizes the glowing *elan vital* of endless development and the imperatives of its sustainability by green capitalism.

The countercultural career of Stewart Brand is instructive inasmuch as he became an uncanny "developmental sustainabler" in both theory and practice. At age 30, in 1968, he launched *The Whole Earth Catalog* to provide the rising ecological counterculture with "access to tools," and, at nearly age 50, in 1987, he co-founded the consulting firm Global Business Network/Monitor to assay how new kinds of electronic "tool access" was altering the Earth. A decade ago, he advanced the idea of "World Earth Discipline" in 2009, having realized "Greens are no longer strictly the defenders of natural systems against the incursions of civilization; now they're the defenders of civilization as well."[17]

Having been in the vanguard of the environmental movement, Brand's response to more radical and popular climate change movements, which radical green thinkers had inspired, was unenthusiastic. For him, their conflicting local, regional, and national agendas often seemed counter-productive and short-sighted. Their efforts had to change, and it was time to become post-environmental, post-ideological, post-philosophical, and post-political by getting more pragmatic. His latest thoughts and projects center on this vague pragmatism, namely, "a practical way of thinking concerned with results rather than with theories and principles."[18] Since the urgent quality of today's climate change crises calls for a "shift deeper than moving from one ideology to another; the shift is to discard ideology entirely."[19] Brand, however, fails entirely at discarding ideology, and his ecopragmatism drifts instead toward ecomodernist solutions. As the proponent of "Whole Earth discipline," Brand in many ways is the epitome of the "ecopragmatic" consciousness behind the ecomodernist challenge, which requires reimagining the relentless resource extractivism at the core of today's global energy regime. To balance out decades of rapidly accelerating growth in hydrocarbon energy use against its known problems, many believe only yet-to-be entirely proven renewable energy alternatives are on the horizon. Brand's endorsement of nuclear energy, however, provides a pretext for inducing an abrupt deceleration in fossil fuel exploitation. During the 1960s, Brand's radical new *Whole Earth Catalogue* lent an aura of faith in new tools, like windmills and solar energy, alone being able to solve any problem. This approach to advancing new technology uses for human survival left him with an aura of tremendous credibility, especially once he joined up with Silicon Valley's emerging cognitive capitalist and informational economy. Are more nuclear reactors, however, truly pragmatic?

Geoengineering, for example, is another option for Brand, and there are many well-framed declarations around its plausibility that allow him to pretend that it can be deployed responsibly. Brand tries to reassure other ecopragmatists with anodyne truisms, noting "one way to geoengineer wrong would be for a private company to start injecting sulfur dioxide into the stratosphere without research plans or results, without outside monitoring of effects, and without permission of a public governance body."[20]

This alibi, however, is incredible, particularly because it implies geoengineers would/could be right in turning to unidentified public monitors or governance bodies to legitimate the process. To mitigate typhoons in the Pacific, for example, might the Rodrigo Duterte or Lee Hsien Loong regimes turn to an Elon Musk, Bill Gates, or Jack Ma to launch geoengineering ventures? While they might not go so far as to propose, like President Trump, a nuclear missile strike on hurricanes, such billionaires could easily procure a research/operation plan with some range of predicted acceptable results to be monitored by other unspecified observers and blessed perhaps by the Earth Governance Network, the UNEP, the Philippine Atmospheric, Geophysical and Astronomical Services Administration, or ASEAN Summit. By Brand's lights, such public-private partnerships for would-be geoengineers might then fairly commission an artificial Pinatubo event or two just to see how such solar radiation management techniques might do the trick intended.

From ecopragmatism to ecomodernism

Often feted as the architects of "an end to 'people are bad' environmentalism," ecomodernists are widely appreciated, on the other end of the rhetorical spectrum, for their new programmatic vision for managing the planet. After 9/11, and with the neo-con cadres at Project for a New American Century (PNAC) promoting their belief that "American leadership is good both for America and for the world," the visions of Schellenberger and Nordhaus would prove seductive.[21] As Eric Holthaus extolled in Slate: "it's inclusive, it's exciting, and it gives environmentalists something to fight for a change."[22]

The "gloom and doom environmentalism" favored by Klein's readers, then, sparked moderate, conservative, or reactionary responses from "New Environmentalism" theorists, like Kevin Kelly, Ted Nordhaus, and Michael Shellenberger. These thinkers argue the "Old Environmentalism" mobilized during the 1960s must retire as this new wave advances its refreshingly more hopeful technocratic vision for finding high-tech paths for ecological modernization programs by creating jobs, slowing climate change, and remaking society. This network of ecopragmatic thought leaders is loose, but fairly extensive and interesting. William McDonough and Michael Braungart, for example, suggest conquering climate change is essentially an aesthetic or design question.[23] If people would only tackle the same big wicked problems from small ordinary angles, like pushing "cradle to cradle design" principles to remake industrial ecologies in the natural economic turnover of Nature's closed loops of energy, matter, and information, things could still be fixed. Seeing prospects for a new green working class and urban renaissance, Van Jones once advanced the idea of creating thousands of new "green collar" jobs in a Green New Deal for the world and its workers, while Chris Lazlo and Paul Hawken embraced new brainstorming through planetarian-scale managerial solutions in new programs for ecological and cultural "corporate social responsibility" at the core of "green capitalism."

Virtually none of these ideas have been ever widely opposed, because the corporate media framing of the recent past and present usually depict these "shades of green" as good, or at least they are not all bad. Moreover, their ecopragmatic presentation of promising green innovations has been so simplistic, unidimensional, and evangelical that they have little room for much serious debate about their premises or promises. McDonough and Braungart perhaps best exemplify the normalization of ecopragmatic outcomes, even though Al Gore, Jr. and

Bill McKibben have also packaged their politics into comforting products adjacent to the ecopragmatists' policy turf, which brings them close to the main salesmen and pragmatic platforms touted by ecomodernism.

The declaration of death for old environmentalism and the ecomodernist turn[24] is now 15 years old, but its ecopragmatist cadres have been intellectually active – if one includes all the political probes made by them with other green capitalist, cradle-to-cradle (c2c) design, urban futurist, bioengineering, and alternative technology thinkers who were allied with them – for arguably 45 years. Obviously, reasoned argument and technological elegance are rarely compelling enough forces to mobilize a movement, enliven the public at large, or alter policy, but there is a striking lack of efficacy with regard to making much material difference for most of these players. Beyond well-feted demonstration projects, admiring glances from afar, and spirited debates when they debut, the impact of ecopragmatism on the planet's environmental crises, even during the more recent years of the Great Acceleration in which the most rapid and pervasive destruction has been inflicted on the environment by burgeoning fossil fuel extraction and utilization, seems somewhat limited.

If the ecomodernists are famous for decrying "People are Bad" environmentalism as being ineffective and paralyzing, one must ask why their push for a strong form of "Technology is Good" environmentalism is proving to be relatively unsuccessful and unpersuasive on their adherents in era of growing environmental awareness? While these groups have assembled an impressive network of authoritative advisors, institutional sponsors, and engaged operatives both in the Washington, DC and the San Francisco metropolitan areas, their policy proposals for fostering deep ecological transformations are not making advances, even during recent years in which big economic, political, and social crises motivated many people to be much more open to opting for grand "shovel ready" big technological solutions to reshape the Anthropocene epoch.

As the *Ecomodernist Manifesto* authors note, "although we have written separately, our views are increasingly discussed as a whole. We call ourselves ecopragmatists or ecomodernists."[25] And, they are not bashful about their foundational assumptions that "The Earth is a human planet," and it "has entered a new geological epoch: The Anthropocene: The Age of Humans."[26] Consequently, their vision is framed around putting humankind's extraordinary powers into the service of "creating a good Anthropocene," which they propose will be non-extractivist, because "natural systems will not, as a general rule, be protected by the expansion of humankind's dependence on them for sustenance and well-being."[27] Here, however, is one of the oddest ruses of reason at work in ecopragmatism. While going on the record to condemn the extractivist expansion of energy, material, and other natural resources as the material cause behind the historic devastations during the Great Acceleration leading into Anthropocene epoch, they also believe those extractivist policies largely await the greater acumen of their "wiser use." Indeed, a refined neo-extractivism remains central to their ecomodernist designs, echoing the same promises that earlier waves of scientific and technological improvement have made in the modernizationist discourses of the last 25, 50, or 100 years.

The qualitative "intensification" of uses for all resources that already are being, have been, or will be extracted, rather than never-ending quantitative "expansion" of conventional extractivism, is the decisive missing link. Intensified, denser, and more concentrated energy, material, and information technology will, in fact, meet the "demand of a good Anthropocene" by bringing ever wiser uses of human "social, economic, and technological powers to make life better for people, stabilize the climate, and protect the natural world."[28] Through these putatively superior logistics and technics, the planet will be saved from mindlessly expanding the material exploitation of Earth's resources. The application of ecopragmatic management will trigger new practices for "intensifying many human activities – particularly farming, energy extraction, forestry, and settlement – so that they use less land and interfere less with the natural world," since this allegedly hitherto undiscovered option "is the key to decoupling human development from environment impacts."[29] Yet uranium mines have a deeply destructive environmental impact that closing coal mines cannot hide.

This "Great Transformation" rests upon creating a new fictive factor of production in capitalist equations; namely, limitless clean energy and wise organization for the intensive "decoupling" of extensive resource use from limitless growth. Pivoting to intensive resource maximization is the core of ecopragmatism, as Brand claims "because we are forced to enter an era of large-scale ecosystem engineering" by those "who work on restoring natural infrastructure."[30] This shift, however, largely turns on the assertion that "the ethical and pragmatic path toward a just and sustainable energy economy requires that human beings transition as rapidly as possible to energy sources that are cheap, clean, dense, and abundant."[31] To reinhabit the Earth, since now, as "a 'city planet' [it] needs city power – grid electricity," Brand also pragmatically asserts, "at present, the best low-carbon source is nuclear."[32] This ecopragmatist proffer is conditional, but it is definitive of their vision for a new vocation, namely, "our obligation to learn planet craft."[33]

In many ways, planet craft is a machinic reboot of Taylorism, mobilizing the managerialist mindset for reducing time, motion, and material inputs into all human activity to amplify time savings, reduce unneeded activity, and minimize material inputs. All of these green Taylorist turns are lumped into the project of "decoupling," because "what decoupling offers is the possibility that humanity's material dependence upon nature might be less destructive."[34] Efficiency rather than excess is the ultimate answer needed to save Nature and rescue Humanity not from, but with capitalism. And, efficient change must opt for "energy technologies that are power dense and capable of scaling to terawatts to power a growing human economy," which means exercising the nuclear option.[35]

Decoupling the industrial economy from extensive exploitation of fossil fuels and other resources at this point in history suggests many forms of less polluting and disruptive "renewable energy[ies] are, unfortunately, incapable" of meeting the bar. Future high-efficiency solar cells, large-capacity batteries, and stored reservoir hydropower innovations will someday qualify.[36] Until then, however, "nuclear fission represents the only present-day zero-carbon technology with the demonstrated ability to meet most, if not all, the energy demand of a modern economy," but even present-day nuclear technologies are too flawed "to achieve significant climate mitigation."[37] As Brand asserts, the calculi of decoupling work in terms of the geographic area sacrificed for power produced as well as waste generated, duration of waste toxicity, scope of waste impact, level of ill health, disability and death experienced, and their cumulative deleterious environmental impact.

For his eco-Taylorist visions of planet craft, Brand mobilizes some rough comparisons to contrast nuclear versus non-nuclear energy for pragmatic ecosystem engineering. An electrical generation plant capable of generating 1,000 megawatts (or 1 billion watts of a gigawatt) for a light water fission reactor takes one-third of a square mile; a wind farm needs 200 square miles, and current PV solar installation requires 50 square miles. High-level nuclear waste for one year at a 1 GW nuclear station in per capita lifetime use of electricity would fill a Coke can, and its yearly consumption of fuel turns 20 tons of fuel into 20 tons of waste that would fill only two dry waste cylindrical storage casks 10 feet in diameter and 18 feet in height. A 1 GW coal plant in per capita lifetime use will produce 68 tons of toxic ash waste and 77 tons of CO_2 as well as heavy metal-infused solid waste; to obtain 1 GW electricity, the plant consumes 3 million tons of fuel, and produces 7 million tons of CO_2 that vented into the atmosphere.[38] In turn, air pollution from coal combustion is also estimated to cause 30,000 deaths a year in the USA, and 350,000 in China,[39] while annual deaths from nuclear energy are more difficult to quantify (absent a catastrophic reactor accident like Chernobyl) due to general background radiation from cosmic rays, solar emissions, geological sources, medical testing, and legacy nuclear weapons testing fallout.

Thus, the ecopragmatism behind ecomodernism reveals the key axiom in its agenda: "nuclear energy is green. Renewables are not green," because they are not "land-sparing, leaving land for Nature."[40] To heed the ecomodernist call for change, ecopragmatism requires "a new generation of nuclear technologies that are safer and cheaper" plus "a conservation politics and a wilderness movement to demand more wild nature for aesthetic and spiritual reasons."[41]

This type of ecomodernist green politics is essentially nuclearizing planetary eco-managerialism. On the one hand, they rightly can credit how a return of "wild nature" and rigid "conservation strictures" now are the norm in heavily protected zones around Pripyat in Ukraine and Fukushima Daiichi in Japan. On the other hand, new ecomodernist groups, like Michael Shellenberger's recently organized Environmental Progress,[42] have ironically pivoted, as green activists, from "save the whales" to "save the reactors," hoping to keep clean nuclear energy generation levels from falling due to early, hasty, or ill-considered decommissioning of nuclear reactors. Therefore, these activists regret that "nuclear comprised 10.5 percent of global electricity in 2016, down from 17.5 percent in 1994," and in the race to match reactor retirements below reactor commissioning/restarts "the world is at moderate to very high risk of losing 203 GW of nuclear energy between today and 2030, and is likely or very likely to add 131 GW of new nuclear by 2030."[43] In addition to the pure energetic efficiency calculations, ecomodernists regard this decoupling as imperative for "aesthetic and spiritual reasons" along with environmental safety concerns, because "next generation solar, advanced nuclear fission, and nuclear fusion represent the most plausible pathways toward the joint goals of climate stabilization and radical decoupling of humans from nature."[44]

There are many ironies here. Eighty years after the first sustained chain reaction in 1939, 70 years after the start of the American and Soviet (now Russian) nuclear arms race, 55 years after the first Chinese nuclear test, and following decades of failed international efforts to regulate nuclear energy, contain nuclear contamination, store nuclear waste permanently, and halt nuclear proliferation, the ecomodernists are full of hope by accepting the ultimate myths of the Atomic Energy Commission and its Cold War-era industrial megamachine to anchor their radical energy transition.

Indeed, green pragmatists opine that "human beings" per se must recognize, although for no stated reason beyond the ecomodernist faith, "such a path will require sustained public support for the development and deployment of clean energy technologies, both within nations and between them, through international collaboration and competition, and within a broader framework for global modernization and development."[45] Their expansive good wishes indeed promise "a rough ride into the future."[46] They also ignore why nuclear fusion experiments, nuclear fission technologies, and nuclear engineering skills are still mostly monopolized within and between more affluent nuclear nation-states, like Britain, China, France, Russia, and the USA, while clean, compact, community-based PV solar arrays, small hydropower turbines, and methane digesters are still dispatched to rural India, Nigeria, Brazil, or Kenya. This imperious neglect of state-based nuclearized arrogance, however, does give greater ideological force to the new political theology unfolding behind the fact-grubbing drive to justify nuclear power, genetic engineering, planetary urbanization, and deep time-keeping to discover a new hermeneutics of the self in the world's exclusive nuclear weapons club in "Whole Earth Discipline." To advance the ecopragmatic government of the self and others, Stewart Brand provides the first commandment for a worldwide ecomodernist culture in his watchwords for ecopragmatism, namely, "We are as gods, and have to get GOOD at it."[47]

To guide his ecopragmatic Genesis, Brand and his decoupling of development and environment within ecomodernism also affirm the neo-Victorian, imperialist, global zoning programs of E. O. Wilson, who proclaims at least "Half-Earth," and especially "the best places in the biosphere" must be put under a strict fundamentalist program for zoological oversight.[48] Its aim is to preserve "wild nature" so that "a great deal of Earth's biodiversity can be saved."[49] Strangely enough, some of the places in question are mostly already under powerful regulation by major G-7 powers, so their conservation by Western nation-states would continue as nationalized "natural resources for the future." Otherwise, many of the other best places for this "Half-Earth" to be created should be carved out of several realms of "The Rest" where "The West" once ruled but does not any longer. Writing an anti-development tract in defense of global conservation, Wilson comes off as yet another designer of a "good Anthropocene." Indeed, his program for planetary preservationism would put much of Mexico, South America, Africa, and Asia in the "Half-Earth" he would reserve for Nature alone without much thought for the Mexicans, South Americans, Africans, and Asians who live there.

Why new and how post-environmental?

To push this ecological program for radical green modernization in the midst of the Great Recession in which there was neither the will nor the way to spend for new jobs, new infrastructure, or new technologies after spending billions to stabilize banks, insurance firms, big companies, and global stock markets seems quixotic. Having declared "the death of environmentalism,"[50] and called on humanity "to love your monsters,"[51] the Breakthrough Institute has worked hard to regulate and direct new post-environmental discursive developments in their own style. In dismissing green global summiteering with their unmoderated carbon control agreements, which still occupy the mushy middle ground of global environmentalism, these thinkers found their mojo again by going "back to the future."

By embracing the Anthropocene, like Stewart Brand, and in pushing for a new age of nuclear energy after the 2011 Fukushima Daiichi nuclear reactor disaster, like Michael Shellenberger, ecomodernists seem to be docking with Buckminster Fuller's "Spaceship Earth" ethics from 50 years ago.[52]

Yet one wonders if these new social forces have become trapped on the same green merry-go-rounds of mid-century modern green thinkers. F. Buckminster Fuller, Paulo Soleri, Paul Goodman, and the Atomic Energy Commission/NRC from the 1950s to the 1970s imagined how to "avoid oblivion" and "attain utopia" for "the whole Earth." Their visions were sweeping, the arguments were sound, the policies were pragmatic. Five decades later, however, no one is living in vast urban megastructures under geodesic domes powered by scores of new nuclear reactors. In a highly disciplined manner, they too pushed for a "whole earth discipline" with denser cities (urban geodesic and archeological megastructures with smaller geographical foot prints), a post-fossil fuel energy regime of nuclear electricity (trading decades of slow decarbonization for rapid radiationization), a "decoupled" environmental protection regime of zoned-off green preserves away from industrial fossil-fueled extractivism (rooted in postindustrial, postmaterialist, and postmodern communities), and intensified anthropogenic life-support systems of reduce/recycle/reuse

Climate change and decarbonization 39

materiality to make Spaceship Earth far more ship shape for human beings setting off to colonize near-earth orbits, the Moon, Mars, and then "to Infinity...and Beyond."

Yet none of this came to pass; it drifted away in the precarity of nuclear détente, the China Syndrome, stagflation, the War on Drugs, national environmental protection acts, and Star Wars ABM theatrics after 1980. The ecoliterati writing in the popular press see this history as a "bad trip." It can be chalked up as a "teaching moment" for the Breakthrough Institute, and the organization receives financial backing from the venture capitalists and new cyber-entrepreneurial foundations thriving in the Greater Silicon Valley/San Francisco Bay area high-tech ecosystem who support private sector space voyages.

It appears that the Breakthrough Institute is striving to regulate modernization less grandiosely. Nonetheless, "post-environmentalism" has been captured by new technology centers, social forces, and vested interests trying to reopen that lost New Frontier to refurbish Earth's habitats by greening nuclear energy, rationalizing planetary urbanization, enforcing technocratic authority, and assuming stewardship of the man-made world of the Great Acceleration, as it was prototyped during the *Trente Glorieuses* of 1945–75. Like the regulatory capture of state agencies, citizen groups, or NGOs by big industries, the rhetorical capture of the Breakthrough Institute supports the aspirations and agendas of other industrial alliances: today's global industrial concerns behind the peculiar social imaginaries of "green nuclearity," "garden megacities," "ubiquitous computing," and greater discipline in adapting to high-tech sustainable degradation beyond fossil fuels with a decarbonized thorium, plutonium, or uranium energy regime. Using ecopragmatism to valorize this vision as "a good Anthropocene," however, does not mean good Anthropocenarians are always going to be at work for the interests of all. As usual, however, one must ask: by who, for whom? Where, when, why, and how?

When such political questions are ignored, theological answers often are provided. In this instance, ecomodernists are providing them. None other than Mark Sagoff, the environmental economist and Breakthrough Institute Senior Fellow, now calls for a theological awakening to adapt people to Brand's deification of Humanity to intentionally revere Gaia by repudiating monotheistic religion. "When God became One, nature became one," and "therefore unitary, singular, and at odds with man, who consumes and corrupts it," which provides the pretext for conservatism as well as doom and gloom environmentalism."[53] For Sagoff, polytheistic religion views "nature as local and assigned gods and spirits to trees, the wind, the harvest, and the Sun," which comports well, in turn, with an ecomodernist pragmatism willing to "understand nature to be many, many places, each with its own guardian spirit. The hope is that human beings will become the guardian spirits of the natural world."[54]

On the one hand, this homily from Sagoff might honor Brand's prophecy that mankind must become like God in the Anthropocene; but, on the other hand, this revelation reminds one of how sequential segments of so many American highways and by-ways are now posted with signage from civic groups – "adopting" two mile stretches of roadways, beaches, or water courses to keep them clean of litter in "We Care" volunteer garbage pick-up campaigns. Usually done at least twice a year, fall and spring, Sagoff's polytheism could be seen as the ecomodern credo for Earth Care doing missionary work for many little local gods. In guarding their denatured anthropocentric turf with true devotion to the local guardian spirits, ecomodernists can enlist in Lions Clubs, Rotarian groups, Soroptimist Clubs, Jaycee groups, Scouting USA troops, or local Costco employee civic action leagues.

The most remarkable revisionism of environmental history, however, is perhaps the ecomodernist understanding for a rewilding of Nature. After knocking the restrictions and regulations of the "old environmentalism," which used rhetorics of ecological destruction, exhaustion, or loss to introduce regulations on fuel economy, energy efficiency, water pollution, landscape devastation, or habitat preservation, ecomodernists at the Breakthrough Institute reduce their collective impact to "a great reversal" in America's use of resources in the 1970s. Naturalizing these efficiencies and reductions as "a series of 'decouplings' … so that our economy no longer advances in tandem with exploitation of land, forests, water, and minerals."[55] Even though historians of energy and material consumption would dispute these claims,[56] ecopragmatists assert America's resources have never been exhausted, but rather "consumers have changed consumption, and because producers changed production."[57]

Yes, producers changed production due to offshoring facilities, and new regulations at home forced technological changes and/or standards alteration. In addition, outsourcing the supplies of energy, materials, and forest products from abroad also changed the levels of pollution from production at home. Similarly, the cadre of "reusing, reducing, and recycling" consumers, who fear for the environment, has altered demand, not due to technological rationalization as much as the secondary effects of regulation plus reframed wants as developed

Climate change and decarbonization 41

by classic environmentalism. These groups also lobbied strongly for more regulatory pressure and legal action. Old-school green politics, not abstract technological change, moved Americans to "spare more resources for the best of nature."[58] And, these limited victories were won by the "old environmentalism" with its catastrophe rhetoric, even though the decoupling trope allows these tangible gains to be ideologically reprocessed as the hitherto immanent unconscious technological innovation driving "post-environmentalism."

Decoupling, energy transition, and new abolitionists

The Breakthrough Institute school of ecopragmatic change in the final analysis asserts: "the ecomodernist project is centrally predicated on decoupling. The only way that 9 or 10 billion people can achieve modern living standards while reducing total environmental impact is if economic growth decouples from environmental impacts."[59] While it claims to have documented this economizing trend in the North Atlantic region, one wonders if the specific intensification of greater particular resources also hides behind less extensive general resource consumption growth.

For example, American agricultural production figures mystify how more or less the same aggregate acreage under cultivation from 1870 to 2010 – 2–4 million acres of farmland – sustained the extraordinary growth in corn output from around 2 million tons in 1870 to nearly 18 million tons in 2010.[60] Looking at constant acreage planted, however, ignores how the exponential rise in nitrogen, potash, phosphates, and water inputs radically increased between 1940 and 1950, and then stayed sustainably high enough to extrude ±15 million more tons of corn produced from the same amount of land with the introduction of new genetically enhanced seed. Citing figures of chemical input reductions in Germany and/or from across the European Union due to that region's higher costs, greater regulation, and different cultivars does not account for American trends. Yet it is mixed rhetorically in the Breakthrough Institute's reports as evidence of today's grand global "decoupling" from expansionist "extractivism" thanks to ecomodernist breakthroughs with intensification.[61]

The simple-minded notion that only expanded acreage under planting equals an extractivist culture crudely occludes how intensifying new chemical, fuel, mechanical, and seed hybrid inputs also advance extractivist outputs with ever-changing intensification measures.[62] To credit "technology" with liberating "the environment" when its manipulated domination only shifts organizationally or technologically from extensive to intensive extractivism is glib. Plainly, a new iteration of "environmentality" is being expressed as a program for fossil fuel decarbonization, which at the same time brings with it one of new nuclear radiationization.[63] When policy interventions are just as decisive as changes in technique, one wonders how transferable these gains are to other ecological settings. Tweaking the tropologies of decoupling is deceptive, since the master trope of extraction stays intact, even as it is out-sourced, off-shored, and over-intensified to downplay expanding resource extraction.

The ecomodernist fixation on nuclear energy also oddly remediates the racialized rhetorics of today's Trumpian times. As Smil, McNeil, Ophuls, and Commoner have suggested in the past, the triumph of fossil fuels in the nineteenth century accelerated the abolition of human enslavement in North America by substituting the energy inputs of dead prehistoric hydrocarbon energy for the work of living carbohydrate-fueled human slaves. These somewhat tortured, but not entirely false, embedded ideologies of individual and collective power were more nakedly conceptualized as what Buckminster Fuller in writings published in the February 1940 issue of *Fortune* magazine on "World Energy" called "energy slaves."[64]

In Fuller's thinking, the inorganic energy slaves were first chained by James Watt in 1776 to coal-fired steam engines, Edwin Drake in 1859 with the first commercial oil wells in Pennsylvania, or Thomas Edison coal-burning boilers at the Pearl Street Station in 1873 with the electrification of modern life in New York City. Fossil fuel use then is directly imagined as a new extractivist order with its own new reserves of "energy slaves," and global oil fields, gas deposits, and coal mines are hybrid formations of power equal to millions of mystified plantations whose inorganic enslaved energies now drive an anthropocentric ethos of greater power, capability, and freedom.

Ironically, Fuller opposed its dirty, noxious, and toxic effects on nature and society with his ephemeral ethic of design that stressed the imperatives of unrelenting efforts to reduce energy and material inputs to minimize entropy and maximize efficiency via elegant design. Even though he did not make this connection, his challenge to the unthinking continuation of wastefully using energy slaves questions the simultaneous dual enslavement/emancipation of many humans by fossil fuels with an aesthetic-technical program for decarbonization with low carbon lifestyles or a post-carbon society. In other words, this vision is the "abolitionist movement" so feared by the Seven Sisters of Oil in the 1970s that saw rapid climate change coming, and then resisted it with their bitter-ender greed to sell each other new and

Climate change and decarbonization 43

improved ICE-powered automobiles, super-efficient gas thermal turbines, or clean coal plants.

Today's true ecomodernists are turning in their crypto-abolitionism to PV panels, windmills, and, most importantly, pressurized light water fission reactors. Nuclear energy for them is the only existing option for sustaining this ideology of freedom, even though a few pretend that "energy, in the form of solar energy, is one economic input that is truly infinite."[65] Solar energy, at this juncture, truly is infinite, but the means for collecting, storing, and applying it are still inefficient, limited, and constrained due to the grids that carry it. It is off in the future, so ecopragmatists stand with Rear Admiral Hyman G. Rickover, the architect of the America's "Nuclear Navy" in the 1950s, who noted:

> With high energy consumption goes a high standard of living. Thus the enormous fossil energy which we in this country control feeds machines which make each of us master of an army of mechanical slaves. Man's muscle power is rated at 35 watts continuously, or one-twentieth horsepower. Machines therefore furnish every American industrial worker with energy equivalent to that of 244 men, while at least 2,000 men push his automobile along the road, and his family is supplied with 33 faithful household helpers. Each locomotive engineer controls energy equivalent to that of 100,000 men; each jet pilot of 700,000 men. Truly, the humblest American enjoys the services of more slaves than were once owned by the richest nobles, and lives better than most ancient kings. In retrospect, and despite wars, revolutions, and disasters, the hundred years just gone by may well seem like a Golden Age.[66]

Rickover's celebratory cantos for carbon, however, were also preludes for the perpetual energy system of a plutonium economy powered by atomic rather than chemical slaves. These metaphors are too good to waste, and the New Abolitionists in the vanguard of today's decarbonization movement already have called this shift to photovoltaic, hydropower, and wind energy the basis of a "The New Abolitionism."[67]

Still, this abolitionism entails another double enslavement: first, of those whose life, liberty, and property are deeply jeopardized by hundreds of new nuclear reactors, uranium mines, reprocessing sites, waste and fuel transport lines, and security threats to come with the radiationization of society; second, of those living without the means, beyond the grid, and lacking the merit to be connected to this putatively new planetary-scale nuclearity supplemented by PV panels, hydropower sites, and wind farms. In 2019, 11 percent of the world's electricity was generated by around 450 nuclear reactors, only increasing this output level to 25 percent would require building over 450 more reactors by 2050 at a rate much greater than the 1980s which had fastest and greater rate of reactor start-ups since the 1950s.[68]

Of course, those with the power, position, and privilege to bask in the nuclear radiance of the world will try to ignore the diktats of ecomodernism, but they are not all likely to succeed in the Breakthrough Institute's post-environmental order given the scale and speed of this imperative new energy extractivism. Their ecomodernist aesthetic essentially flips NIMBYism on its head, suggesting that the ways "environmentalists valorize NIMBYism in general as a kind of authentic and democratic placed-based activism" is, in fact, an unjust irrational prejudice.[69] Arguing against the affluent limousine liberal residents of Nantucket Island who protested against installation of a large industrial-scale Cape Wind windmill farm in Nantucket Sound, ecopragmatist reasoning plays another "race card," suggesting such site defense actions are greenlining their residences, neighborhoods, or localities out of irrational fear, class status, and even racial privilege to avoid seeing these energy plantations. Indeed, "justifying one's politics according to one's *place* is as much a prejudice – and often as much a strategy to protect privilege – as justifying one's politics according to *race*"; therefore, they twist equal opportunity multiracial amity in any interaction to valorize the fungibility of all places for any use – if we do not "confine our affections to or reserve our loyalties for a particular race. Why then do we believe we are justified in reserving our loyalties for a particular place?"[70]

Such positions in ecomodernist reasoning are static, conservative, undemocratic, elitist ideological mystifications inasmuch as they are "inherently averse to change," especially given how "science tells us that almost all places – from the shifting continents to deserts that were once oceans to ancient cities that are literally built on settlements that came before them – are in continual change and evolution."[71] Privileged ecoconservatism and progressive ecomodernism ultimately do not share the same epistemic or aesthetic sensibility – "when we look a windmill, a bridge or building, do we really see the same fact? Are the windmills ugly or beautiful? Are they symbols of our degradation of nature – or are they symbols of a brighter future?"[72] To decouple society from fossil fuels, dirty hydrocarbon wastes, and ugly carbon infrastructure, ecopragmatists trump ordinary residents, first,

Climate change and decarbonization   45

with their technocratic rigid green threats, e.g., "Cape Cod, for example, *without* a wind farm but with oil tankers."[73] And, then, second, the rapture of a new technological sublime, which is supposedly dynamic, progressive, equitable, and popular; indeed, it can be mobilized, because so

> many observers have commented that they find the blades, which rotate languidly in even the fiercest gales, hypnotic. Tourists stare at them, reveling in their whisper-quiet power. Some people compare them to peaceful giants or scarecrows. But to our mind, the most delicious comparison is with those of other wind-powered devices that grace the waters around Cape Cod: sailboats.[74]

By rejecting "the prejudice of NIMBY place," then, ecomodernism is free to reshape the supremacy of their ultra-technified spaces and pragmatically organized energy slave enclaves. So too will the cooling towers, reactor domes, and reprocessing plants of their intensified and expansive nuclear economy perhaps be likened to the temples and monuments of Rome in their bondage to the enslavement of nuclear energy slaves after the total, or nearly complete, abolition of fossil fuels.

Conclusion

Such counter-intuitive admirations for high-tech sustainable development programs are often only implicit or tacit in ecopragmatic discourse, but they are very real. In 2008, when Van Jones and others initially pushed a "Green New Deal" in the first Obama campaign for the White House, Thomas Friedman expressed the inherently conservative core of sustainability theories and practices clad in the rhetoric of innovative change for the USA in his visions for averting "a hot, flat crowded world" in the twenty-first century:

> The era we are entering will be one of enormous social, political, and economic change – driven in large part from above, from the sky, from Mother Nature. If we want things to stay the same as they are – that is, if we want to maintain our technological, economic, and moral leadership and a habitable planet, rich with flora and fauna, leopards and lions, and human communities that can grow in a sustainable way – things have to change around here, and fast.[75]

Amidst the Great Recession, Friedman asserted this change was coming as "innovators and idealists' related to him "new ideas for making clean energy" and "new thoughts about how to repair something in our country that desperately needs repairing."[76] Some of those ideas persist; but, during the intervening years of financial crisis, populist backlash, geopolitical drift, and ongoing war, American elites and mass publics have become more distracted by what Friedman tagged as "dumb as we want to be" politics rooted in red state/blue state divisions, mindless nationalism, and aggressive isolationism.

As the days of Sarah Palin slipped away into years of Donald Trump, many resented and resisted the efforts made by the nation's first African-American president to highlight how "changes in our politics and temperament over the last three decades – not only after 9/11 – have fractured our focus and privatized our will."[77] In this vacuum of neoliberal distraction, institutionalized greed, and nationalistic privilege, which began in 1980 with President Reagan's moves toward illiberal populist democracy, things are worse. The more stressed conditions of 2019 are pushing ecomodernists and others to overemphasize illiberal elitist technocracy to advance their new designs for reengineering everyday life from above in the name of "sustainable development" under the banners of ecomodernism.[78]

Once again, the tacit elitism in ecomodernist planetarian accountancy, management, and operations shows itself in ecopragmatist sustainability reasoning.[79] Shellenberger's and Nordhaus's narrative ploys in 2019 are not unlike Commoner's or Fuller's in 1969: how to avoid, or at least forestall as long as possible, widespread ecological collapse within a global economy predicated upon stimulating endless material growth. Given how the avowed goals of the Breakthrough Institute are "ecopragmatic," their wager on nuclear energy, denser urbanism, transgenic agriculture, and geoengineering affairs aim directly at assuring the supply of stable, secure, and sustained positional goods for the core G-8 gaggles of nation-states where effective governmentality over territory and population still prevails.

The Breakthrough Institute's multiple technoliberal strategies for avoiding ecological collapse fulfill the requirements of what Baudrillard asserts is "the casual événementalité of our age... which is that of deterrence, the baleful form of which presides over the nullity of our age."[80] The greatly feared antithesis of sustainability, which environmental thinkers like Diamond have exclaimed, is "collapse."[81] Ecological "overshoot," economic "unsustainability," and environmental "degradation" are, by many measures, occurring, but their incidence and scope generally still are local or regional. The conditions

Climate change and decarbonization 47

of continuous, if not complete, ecological destruction are not yet truly at the tipping point for extensive planet-wide collapse. Hence, the ecomodernist faith in nuclear fission, GMO-based agriculture, and planetary urbanization as the culmination of "sustainable development" can be compared to "strategic deterrence" in the entirely ironic manner that Baudrillard discloses it. That is,

> deterrence is a very peculiar form of action: it is *what causes something not to take place*. It dominates the whole of our contemporary period, which tends not so much to produce events as to cause something not to occur, while looking as though it is a historical event.[82]

The emptiness of the Breakthrough Institute's environmental summits, nuclear reactor rescues, and ecomodernist development programs, as well as yet another not-for-profit green organization trying to trigger new social movements, become much more obvious by tracing out this *Zeitgeist* of deterrence: "it is a diabolical force which wrecks the actual acting out of events or, if they still take place–if they have taken place–destroys their credibility."[83] Scores of ecopragmatic sustainability conferences – 15 years after "the death of environmentalism" – have either not developed much or sustained even less after Shellenberger and Nordhaus invited everyone "to love your monsters" at the coming of the Anthropocene. Even though his reading of such outcomes is extreme, Baudrillard seems correct about the activities undertaken to prevent some ultimate ecological collapse yet to be experienced. It is not the lost Holocene or "a void left by the ebbing of past events, but a void due to the sucking effect of a future event ... creating a violent depressurizing of the social, political, cultural and mental sphere," which is coming from the Anthropocene.[84] What fills the vacuum, even as it fights to maintain the conceptual pressurization of the green public sphere, is EnvironmentalProgress. org clanging its tocsin to keep existing nuclear reactors from becoming extinct and support building hundreds more.

These apparently authentic victories for alternative energy, radical design, appropriate technology, and collaborative governance could keep many communities in the Global South frozen in time or marching in place. The poor may gain a better built environment, their NGO assistance networks might help them maintain it, and the ecological subversives fighting for sustainability will try to ensure their advancements are sustained. Nevertheless, reactionary edges are concealed in such ecological achievements inasmuch as they enforce an eco-managerial planetarian regime that sells "development as freedom" and "sustainability as security" by getting more bicycles, rooftop home PV panels, village-scale windmills, and reclaimed plastic building blocks to upgrade the Global South's half-planet of ecopragmatic shanty-towns.

In its fullest ecomodernist forms, hi-tech entrepreneurs, corporate oligarchs, or new class elitists, like Stewart Brand, hedge their bets by paying homage to "intentional Gaia." Such Earth worship makes it apparent they aim to control who gets what, where, when, and how from the environment. Whatever scarcity develops for any natural resource market, and the economic losses it might entail, will be shifted to others down and out below, quickly and permanently. Meanwhile, relative abundance can still be enjoyed by the more planetarian-minded cadres of the Breakthrough Institute and others like them. The restored wildlands in nature preserves on our "blue green planet" are, in part, for the biosphere, but the richest range of opportunities this well-preserved biosphere brings will be kept open, first and foremost, to green globalists and their allies whose basic tactics shield such sites and their ecological services from the hordes of people inhabiting the planet of slums.

The "post-environmentalism" of the Breakthrough Institute's senior managers and fellows essentially touts another grand revision in energy extractivism. Their simple formulae for "a good Anthropocene" require nuclear electrification, urban densification, and genetic engineering plus the planetarian eco-managerialism of "Whole Earth Discipline," curating Wilson's rewilded "Half-Earth" decoupled from remainder of the globe that will be "City Planet." Brand recasts this change as the historic destiny of the human race. Even though which humans have a deeper designation for greater destiny is left unanswered, ecopragmatists are certain: "whether it's called managing the commons, natural-infrastructure maintenance, tending the wild, niche construction, ecosystem engineering, mega-gardening, or intentional Gaia, humanity is now stuck with a planet stewardship role."[85] Not all humans, however, will be able to enroll as stewards for the planet.

More pointedly, are there political dangers in granting trust blindly those who identify supposedly "commonsensical opportunities" to transform American environmental politics with new pragmatic policies? Granted, how the ecomodernists seem increasingly at best to be the vanguard intelligentsia of new type of nuclear technoliberalism to contain, control, and constitute another energy extractivism beyond deep carbonization, should they be trusted center their decarbonizing campaigns on ramping up a massive radiationization of the energy regime by deploying dangerous shunned technologies from the Cold War era? Atmospheric CO_2 levels were 374 ppm in October 2004; they are close to 411 ppm in March 2019.[86] Despite a faith in the power of science and rationally discovered evidence presented by scientists like James Hansen in 1988, it is ominous that "833 GtCO2e [gigatons of carbon dioxide emissions] was emitted in just 28 years since 1988, compared with 820 GtCO2e [gigatons of carbon dioxide emissions] in the 237 years between 1988 and the birth of the industrial revolution."[87] After declaring "death of environmentalism" in 2004, this trend does not represent a radical rejection of energy extractivism. Nor does it promise to slow the global warming that an ecopragmatic breakthrough was promising 10, 15, or 20 years ago, especially given how nuclear energy generating capacity in the USA is less in 2019 than it was during the mid-1990s.

To conclude, this analysis disputes the merits of an ecomodernist program for new green economies and societies. First, ecopragmatist programs of change have outlined grand plans for how to move the world's nations ahead in an expedient green capitalist manner, but without having much material effect. And, second, ecomodernists have articulated what they regard as a devastating critique of earlier environmentalist thought and practice, but they do not demonstrate why they are fatally flawed failures. Ecomodernist solutions also have had little tangible success, even though they keep on spinning up positive reviews to sustain their professionally correct pragmatic attitudes and ethos. Finally, this study suggests that ecopragmatic designs for the radiationization of economies and societies to decarbonize them would only transmutate the extent, intensity, and type of energy extractivist practices. The final outcome is a strange politics of delusion, delay, and destruction in ecomodernism's impact and implementation that is opening black holes in this cosmic vision of our green future.

Notes

1 Jerry K. Jacka, "The Anthropology of Mining: The Social and Environmental Impacts of Resource Extraction in the Mineral Age," *Annual Review of Anthropology* 47, no. 1 (2018): 61–77, doi:10.1146/annurev-anthro-102317-050156.
2 Oliver Milman, "Ex-Nasa Scientist: 30 Years on, World Is Failing 'Miserably' to Address Climate Change," *The Guardian*, June 19, 2018, sec. Environment,www.theguardian.com/environment/2018/jun/19/james-hansen-nasa-scientist-climate-change-warning.

3 Milman.
4 Milman.
5 Nathaniel Rich, *Losing Earth: The Decade We Could Have Stopped Climate Change* (London, UK: Picador, 2019).
6 Paul Hawken, Amory Lovins, and L. Hunter Lovins, *Natural Capitalism: Creating the Next Industrial Revolution* (New York, NY: Little, Brown & Company, 1999); Thomas L. Friedman, *Hot, Flat, and Crowded: Why We Need a Green Revolution – And How It Can Renew America* (New York, NY: Farrar, Straus and Giroux, 2008); John Elkington, *Cannibals with Forks: Triple Bottom Line of 21st Century Business* (Oxford, UK: John Wiley & Son, 1999).
7 Ted Nordhaus and Michael Shellenberger, "The Death of Environmentalism: Global Warming Politics in a Post-Environmental World," *Grist* (blog), 2004, https://grist.org/article/doe-reprint/.
8 "Programs | EGA," accessed November 25, 2019, https://ega.org/about/programs.
9 Michael Shellenberger and Ted Nordhaus, eds., *Love Your Monsters: Postenvironmentalism and the Anthropocene* (San Francisco, CA: Breakthrough Institute, 2011).
10 Ted Nordhaus and Michael Shellenberger, *Break Through: From the Death of Environmentalism to the Politics of Possibility* (New York, NY: Houghton Mifflin, 2007).
11 "Breakthrough Dialogue 2019: Whole Earth Discipline," The Breakthrough Institute, accessed October 19, 2019, https://thebreakthrough.org/events/breakthrough-dialogue-2019-whole-earth-discipline.
12 Nordhaus and Shellenberger, "The Death of Environmentalism."
13 Naomi Klein, *The Shock Doctrine: The Rise of Disaster Capitalism* (New York, NY: Metropolitan Books, 2007); Naomi Klein, *This Changes Everything: Capitalism vs. the Climate* (New York, NY: Simon & Schuster, 2014).
14 Peter Huber, *Hard Green: Saving the Environment from the Environmentalists: A Conservative Manifesto* (New York, NY: Basic Books, 1999).
15 Stewart Brand, *Whole Earth Discipline: Why Dense Cities, Nuclear Power, Transgenic Crops, Restored Wildlands, and Geoengineering Are Necessary* (New York, NY: Penguin Books, 2010).
16 Elkington, *Cannibals with Forks*.
17 Brand, *Whole Earth Discipline*, 1.
18 Brand, 1.
19 Brand, 1.
20 Brand, 311.
21 "Project for a New American Century: PNAC Statement of Principles," 1997, www.rrojasdatabank.info/pfpc/PNAC – statement%20of%20principles.pdf.
22 Eric Holthaus, "Manifesto Calls for an End to 'People Are Bad' Environmentalism," *Slate Magazine*, April 20, 2015, https://slate.com/technology/2015/04/ecomodernism-a-21st-century-environmental-philosophy-that-embraces-a-good-anthropocene.html.
23 William McDonough and Michael Braungart, *The Upcycle: Beyond Sustainability – Designing for Abundance* (New York, NY: North Point Press, 2013); William McDonough and Michael Braungart, *Cradle to*

Cradle: Remaking the Way We Make Things (New York, NY: Farrar, Straus and Giroux, 2010).
24 Nordhaus and Shellenberger, "The Death of Environmentalism."
25 John Asafu-Adjaye, "An Ecomodernist Manifesto," April 2015, 7, ecomodernism.org.
26 Asafu-Adjaye, "An Ecomodernist Manifesto."
27 Asafu-Adjaye, 6.
28 Asafu-Adjaye, 7.
29 Asafu-Adjaye, 7.
30 Brand, *Whole Earth Discipline*, 23.
31 Asafu-Adjaye, "An Ecomodernist Manifesto," 25.
32 Brand, *Whole Earth Discipline*, 23.
33 Brand, 23.
34 Asafu-Adjaye, "An Ecomodernist Manifesto," 25.
35 Asafu-Adjaye, 22.
36 Asafu-Adjaye, 23.
37 Asafu-Adjaye, 23.
38 Brand, *Whole Earth Discipline*, 80–81.
39 Brand, 82.
40 Ausubel cited in Brand, 86.
41 Asafu-Adjaye, "An Ecomodernist Manifesto," 23, 27.
42 "Clean Energy Crisis," Environmental Progress, accessed November 25, 2019, http://environmentalprogress.org/global-crisis.
43 "Clean Energy Crisis."
44 Asafu-Adjaye, "An Ecomodernist Manifesto," 23–24.
45 Asafu-Adjaye, 24.
46 James Lovelock, *A Rough Ride to the Future* (New York, NY: The Overlook Press, 2015).
47 Brand, *Whole Earth Discipline*, 1.
48 Edward O. Wilson, *Half-Earth: Our Planet's Fight for Life* (New York, NY: Liveright, 2016), 133–53.
49 Wilson, 136.
50 Nordhaus and Shellenberger, *Break Through*.
51 Shellenberger and Nordhaus, eds., *Love Your Monsters*.
52 R. Buckminster Fuller, *Operating Manual for Spaceship Earth* (Carbondale: Southern Illinois University Press, 1970); Sabine Höhler, *Spaceship Earth in the Environmental Age, 1960–1990*, History and Philosophy of Technoscience 4 (London, UK: Routledge, 2015).
53 Mark Sagoff, "A Theology for Ecomodernism," *Breakthrough Journal* 5 (2015), https://thebreakthrough.org/journal/issue-5/a-theology-for-ecomodernism.
54 Sagoff.
55 Jesse H. Ausubel, "The Return of Nature," *Breakthrough Journal* 5, accessed November 25, 2019, https://thebreakthrough.org/journal/issue-5/the-return-of-nature.
56 J. R. McNeill and Peter Engelke, *The Great Acceleration: An Environmental History of the Anthropocene since 1945* (Cambridge, MA: Belknap Press, 2016); Vaclav Smil, *Energy and Civilization: A History* (Cambridge, MA: MIT Press, 2017); Vaclav Smil, *Power Density: A Key to Understanding Energy Sources and Uses* (Cambridge, MA: The MIT Press, 2016);

Vaclav Smil, *Energy Transitions: Global and National Perspectives*, 2nd Edition (Santa Barbara, CA: Praeger, 2010); Fairfield Osborn, *Our Plundered Planet* (Boston, MA: Little, Brown and Company, 1948); William Vogt, *Road To Survival* (New York, NY: William Sloane Associates, 1948).
57 Ausubel, "The Return of Nature."
58 Ausubel.
59 "Breakthrough Dialogue 2019: Whole Earth Discipline."
60 Ausubel, "The Return of Nature."
61 Ausubel.
62 Ausubel.
63 Timothy W. Luke, "On Environmentality: Geo-Power and Eco-Knowledge in the Discourses of Contemporary Environmentalism," *Cultural Critique*, 31 (1995): 57–81, doi:10.2307/1354445.
64 F. Buckminster Fuller, "Trends and Transformations," *Fortune*, February 1940.
65 Christian Parenti, "If We Fail," Jacobin, August 29, 2017, https://jacobinmag.com/2017/08/if-we-fail.
66 Hyman Rickover, "Energy Resources and Our Future" (Annual Assembly of the Minnesota State Medical Association, St. Paul, MN, 1957), 12, http://large.stanford.edu/courses/2011/ph240/klein1/docs/rickover.pdf.
67 Chris Hayes, *The Nation*, "The New Abolitionism," May 12, 2014, www.thenation.com/article/new-abolitionism/.
68 "Nuclear Power Today | Nuclear Energy – World Nuclear Association," accessed November 25, 2019, www.world-nuclear.org/information-library/current-and-future-generation/nuclear-power-in-the-world-today.aspx.
69 Nordhaus and Shellenberger, *Break Through*, 100.
70 Nordhaus and Shellenberger, 101.
71 Nordhaus and Shellenberger, 102–3.
72 Nordhaus and Shellenberger, 103–4.
73 Nordhaus and Shellenberger, 102.
74 Nordhaus and Shellenberger, 104.
75 Friedman, *Hot, Flat, and Crowded*, 7.
76 Friedman, 9.
77 Friedman, 11.
78 Al Gore, *The Future: Six Drivers of Global Change* (New York, NY: Random House, 2013).
79 Timothy W. Luke, "Developing Planetarian Accountancy: Fabricating Nature as Stock, Service, and System for Green Governmentality," in *Nature, Knowledge and Negation*, ed. Harry F. Dahms, vol. 26, Current Perspectives in Social Theory (Emerald Group Publishing Limited, 2009), 129–39, doi:10.1108/S0278-1204(2009)0000026008.
80 Jean Baudrillard, *Jean Baudrillard: Selected Writings: Second Edition*, ed. Mark Poster, trans. Jacques Mourrain (Stanford, CA: Stanford University Press, 2002), 256.
81 Jared Diamond, *Collapse: How Societies Choose to Fail or Succeed*, Revised edition (New York, NY: Penguin Books, 2011).
82 Baudrillard, *Selected Writings*, 256.
83 Baudrillard, 257.
84 Baudrillard, 257.
85 Brand, *Whole Earth Discipline*, 275.

86 "Monthly CO_2," accessed November 25, 2019, www.co2.earth/monthly-co2.
87 Paul Griffin, "The Carbon Majors Database: CDP Carbon Majors Report" (CDP, July 2017), 7, https://b8f65cb373b1b7b15feb-c70d8ead6ced550b4d987d7c03fcdd1d.ssl.cf3.rackcdn.com/cms/reports/documents/000/002/327/original/Carbon-Majors-Report-2017.pdf.

4 Star power

Outer space mining and the metabolic rift

Emily Ray and Sean Parson

Introduction

Every day, about 100 tons of interplanetary material rains down on Earth, mostly as dust from ancient cosmic collisions.[1] Although likely few are aware of this constant contact with outer space materials, this statistic serves as a reminder that the boundary between outer space and earthly lives is porous and always open to the extraterrestrial. The most well-known photograph of Earth, taken from space in 1972, was immediately appropriated by the environmental movement of the 1970s to project the vulnerability of the "blue marble" that is our only home planet. The National Aeronautics and Space Administration (NASA) Center for Near Earth Object Studies website offers a primer on Near Earth Objects (NEOs) and Near Earth Asteroids (NEAs). NEOs include comets and asteroids of a certain minimum size, and the majority of NEOs are asteroids. Given the general interest in both, we will use NEOs and NEAs interchangeably throughout this chapter. One tab in the online primer, "Target Earth," describes a need for "some form of NEO insurance" to protect against an eventual devastating hit from an NEO; another reminder of the planet's frailty. But the next tab over, "NEAs as Resources," projects an entirely different relationship with Near Earth Asteroids that just before spelled potential doom for life on Earth. "The comets and asteroids that are potentially the most hazardous because they can closely approach the Earth are also the objects that could be most easily exploited for their raw materials."[2] NASA thus presents a compelling new way to relate to NEOs and NEAs. Rather than encounter them as objects which must be kept at a distance, one ought to encounter them as opportunities for resource extraction; opportunities only made available by their dangerous proximity to the planet. This geospatial tension between "close enough" and "too close" is a good approximation of the tension in the continued commitment to extractive industries in a climate radically changed by them. For many wealthy entrepreneurs, NEOs and NEAs are "close enough" to be made exploitable, and the potential extractive resources on these objects can extend the life of the terrestrial economic system to continue to produce wealth, even as it pushes "too close" to ending an incredible run of multi-species flourishing on the planet.

This chapter critically explores the growing legal and policy framework put into place for extraterrestrial private mining and resource extraction in the United States. At present, the purpose of investing in space exploration and asteroid mining is: to bring down the cost of space travel by using the resources available on asteroids to support space missions and colonies, to make scientific discoveries about Earth through the study of nearby celestial bodies, and to explore creative responses to climate change by engineering Earth-saving technologies located in outer space. The ability to act on these purposes is guided by a handful of outer space policies meant to avoid disastrous competition between nations. These policies, when combined with the investment and speculative interest by private industry, are part of a larger process of neoliberalizing outer space, particularly through public-private partnerships. While these policies attempt to commodify NEAs, they do significantly more than just that. They are part of a project to redefine the boundaries of the planetary – which also provides a means for capital to circumvent ecological resource limits – while providing a psychological focus on speculative technologies in order to obfuscate debates around the restructuring of economic, social, and political institutions on Earth. Lastly, even if the projects are economically successful in the short term – and there is no economic guarantee of that – the resulting impact would be catastrophic for Earth's ecological systems. Using Foster, York, and Clark's work on the *Ecological Rift*, and O'Connor's work on eco-Marxism, we contend that the introduction of consumptive resources from outside the terrestrial ecological system will do nothing but further exacerbate the ecological crisis.

The ecological rift and ecological catastrophe

Asteroids are "rocky, airless remnants left over from the early formation of our solar system about 4.6 billion years ago" that orbit around the sun.[3] At the most recent count, there are 794,145 asteroids, mostly between Mars and Jupiter. There are three classifications for asteroids, each indicating the materials it is composed of. Asteroids typically contain metals that have great economic value on Earth, including platinum, iron, and nickel. Some contain water, which could be used to produce rocket propellent and to support colonization missions.[4] NASA, and the range of private companies vying for the chance to support experimental asteroid mining missions, describe neither asteroids nor other celestial bodies as being part of nature or the environment. Critical ecological perspectives, which reject the pursuit of outer space mining and colonization, argue for putting energy into salvaging Earth instead. From this perspective, outer space exploration is a frivolous use of public funds. However, outer space is unique in that there are virtually no concerns about the loss of outer space environmental integrity in the same way that John Muir (the Sierra Club) and Howard Zahniser (the Wilderness Society) passionately rearticulated the American conception of wilderness from a hindrance to, or symbolic of, progress, to being sacred in the American imaginary. NEAs are fairly easy to categorize as a "gift" of resources, recalling what Marx described as the wrong-headed view of nature as inherently exploitable through the consumption of resources in a capitalist economy.

James O'Connor has persuasively argued that there are two contradictions of capitalism. "The first contradiction of capitalism is internal to the system; it has nothing to do with the conditions of production...," which leaves the second contradiction centered on the "...economically self-destructive appropriation and use of labor-power, urban infrastructure and space, and external nature or environment."[5] One of the responses to the many crises of capitalism is for the state to intervene to encourage extractive and productive uses of the environment: "...every state agency and political party agenda may be regarded as a kind of interface between capital and nature..."[6] The Outer Space Treaty (OST) can be viewed as an example of state intervention to manage the potential commercialization of outer space and to manage the exploitation of the resource needs of individual nations. The OST, Moon Treaty, and Commercial Space Launch Competitiveness Act (Space Act) present a transition from a global approach to managing outer space as a set of economic resources to, so far, the most supportive of liberating capital into the heavens. "Although the capitalization of nature implies the increased penetration of capital into the conditions of production...the state places itself (or mediates) between capital and nature, with the immediate result that the conditions of capitalist production and politicized."[7] An important point to take from this is that despite the efforts of space entrepreneurs to frame their investments as humanitarian and economic endeavors, theirs is an explicitly political project, too. Neoliberalism re-presents political actions as apolitical via a narrative of simply applying technical knowledge into profitable practice available to anyone in a globalized capitalist economy where anyone can pay to play, and those who cannot will benefit from the risks taken by investors and innovators. Asteroid mining is thus presented simply as a matter of funding the technological advancements to make possible the extraction of resources for space colonization and, possibly, for the transport of mineral wealth back to Earth (although this is considered a harder project to economically justify than using asteroids to support space expeditions). Investors need the state to negotiate the terms of outer space resource extraction and use so their efforts can be rewarded with publicly funded contracts from agencies like NASA in order to monetize the work that they claim no one else will do, but it is necessary to rescue the human species from the inevitable death of human consciousness on an increasingly climatologically volatile Earth.

Similar efforts to commoditize terrestrial natural resources as necessary conditions of capitalist production have been met with some resistance from various social and environmental groups. Activism ranges widely from major mainstream organizations like the National Resource Defense Council to local, indigenous direct action resistance movements. However, asteroids and celestial bodies have yet to stir the environmental movement to include outer space as part of a defendable environment that has value beyond serving as the base of commodity production. While this chapter does not argue for or against the exploration of space or use of asteroids to support commodity production, it does insist on a critical analysis of this move in order to draw attention to the extension of capitalism into the cosmos, supported by a neoliberal framework for state and non-state actors.

The promise of outer space resource extraction is attractive even to those who do not promote a capitalist vision of the future. In his book, *Fully Automated Luxury Space Communism*, Aaron Bastani makes an argument for a communist politics that rejects the logic of scarcity and instead imagines a post-scarcity, post-work world.[8] In this vision, the process and practice of labor would be mostly automated and, for the first time in human history, the full range of luxury – which is currently only enjoyed by the rich – would be democratized. Bastani, to his credit, does not see the possibility of a post-scarcity world without radically transforming current energy systems, nor without extracting materials in space. Since climate change and broader environmental destruction from extractive capitalism have brought the planet to the ecological breaking point, Bastani understands that a luxury, growth-oriented, post-scarcity politics is not possible on Earth with the purely terrestrially grounded technology of the current context. He writes, in a clear op-ed for the *New York Times*,

> And yet it is true: Ours is an age of crisis. We inhabit a world of low growth, low productivity and low wages, of climate breakdown and the collapse of democratic politics. A world where billions, mostly in the global south, live in poverty. A world defined by inequality.[9]

To overcome this shortcoming, he looks to the opportunities in space asteroid mining:

> More speculatively, asteroid mining — whose technical barriers are presently being surmounted — could provide us with not only more energy than we can ever imagine but also more iron, gold, platinum and nickel. Resource scarcity would be a thing of the past.

He claims that by mining resources in space, there is no need to despoil Earth ecosystems, instead injecting the resources needed to maintain high levels of personal consumption into the earthly industrial metabolism. An uncritical reading of this work, which dovetails with the visions of Silicon Valley and the neoliberal space industry, might seem convincing. This is because Bastani and Silicon Valley understand pollution only as a problem of production. They do not, as Marx did, understand the much more complex metabolic relationship that exists between production, consumption, and disposal.

The work that best explores the dialectical metabolic relationship is Foster, York, and Clark's work on the metabolic rift, which is useful for thinking through this attempt to loop outer space into the metabolic relationship between production and the environment.

> For Marx, the metabolic rift—the alienated mediation between humanity and nature—was a product of the 'robbing' or expropriation of the soil, and thus of nature, thereby hindering the operation of the eternal natural condition for the lasting fertility of the soil.[10]

Here, Marx argues that ecological crisis is, in part, due to the geographic separation of production and consumption. Capitalism takes massive amounts of resources and nutrients from the land and concentrates them in dumps and urban areas. The metabolic rift is also a product of urbanization and increase in population density. The ecological rift shatters the relationship within a local environment by creating an imbalance. Either there are too many nutrients being taken from the land or too much material being added to it. That fissure is the ecological crisis. The ecological crisis is not simply about inputs or outputs, about extraction or production, but is instead an issue of relationality.

Marx also argued that humans are innately driven to produce; that there is a need to creatively use that capacity for labor, and this accounts for a distinction between humanity and other animals. Non-use of the environment is therefore not an alternative to the ruptured metabolic rift. Rather, non-alienated use of the environment, governed by a different set of values than those of neoliberal capitalism, can provide the foundation for a sustainable, just, and responsible relation between human productive energy and the foundation of production in the form of the environment. Steven Vogel has similarly argued that the problem is not *that* humans engage the environment for resource use, but *how* they do so that is problematic.[11] Is it possible to see the cold, hard rocks floating in a freezing vacuum as part of the environment that requires a new, non-capitalist approach to managing the conditions for human life? These same rocks are already considered part of the available environmental resources for Earth-based needs, and including them in the movement of the metabolism on Earth, or even off-Earth but in conjunction with on-Earth resources, such as space exploration missions, only answers the question of their place in the environment from the perspective of capital. As the private sector gains more control over the governing of these NEAs, the task of social theory is to determine if, and how, outer space should be ideological and practically linked into the metabolic relationship between production and the sustenance of life.

The process of taking resources from the extra-planetary environments of asteroids would only further exacerbate this crisis. One of the primary narratives of contemporary capitalism is around resource scarcity, which claims that the limits of human advancement are linked to the amount of resources on this planet. Defenders of such practices argue that adding resources from an outside system would address the crisis of environmental degradation from terrestrial resource extraction. Looking to the stars to address the issue of resources and growth is simply a means to circumvent political discussions around resource allocations and radical shifts to human, economic, and social systems that are conducive to human survival. In order to avoid addressing the contradictions of capitalism, or the ecological limits that bind the human species, these arguments require magical thinking and grandiose technological innovation. This sleight of hand distracts from an analysis of the root problem of the crisis by focusing on scarcity – which Marcuse decades ago had already shown is illusory.[12] As Marcuse shows, the problem is not one of scarcity but one of over-abundance. Adding additional resources into the closed system that is the ecosystem would exacerbate the metabolic rift, not solve it. Looping NEOs into the industrial metabolism not only adds additional materials to an already overburdened planetary metabolism but also overlooks the troubling relationship between neoliberalism and outer space, reconceived as a frontier for capital expansion, and fraught with difficult questions about national sovereignty. What is at stake in extending the planet beyond Earth itself?

Defining the limits of the planetary

The process of defining and categorizing the world, as Foucault persuasively argues in *Archeology of Knowledge*, *The Order of Things*, and *Discipline and Punish*, is an inherently political process. The construction of a category is a conduit of power, a form of action that also creates narrative limits and restrictions that are central to structuring society. While science is often considered beyond the scope of politics and social hierarchies of power, this is not the case. In fact, the fields of science and knowledge are central node in the networks of power. Science studies scholar Donna Haraway, in *Modest_Witness@ Second_Millenium.FemaleMan_Meets_Onco_Mouse*, argues that the social relations of a society are inherently involved in the construction of scientific concepts and "laws." In effect, science is inherently a social force and the products of science are thus both social construction and entities that structure and regulate the social realm. This is true in biology, when discussing concepts like species or race, and is also the case when it comes to defining what is, or is not, part of the planet. Kathryn Yusoff illustrates a similarly political process in *A Billion Black Anthropocenes or None* when they explore the complex racial and gendered development of the field of geological sciences.[13] While it is easy to assume that there is a clear dividing line separating the planetary from the non-planetary, in reality that line, like all borders, is hard to define, porous, or nonexistent. Defining the planetary is, like all other categorization, a political act.

Redefining the planetary requires thinking about both borders and frontiers.

All nations have borders, and many even today have walls. But only the United States has had a frontier that has served as a proxy for liberation, synonymous with the possibilities and promises of modern life and held out as a model for the rest of the world to emulate.[14]

Grandin argues that the frontier myth did not die even after the United States pushed colonization and expansion out to the Pacific coast. A border suggests a closing: a hard boundary between friends and enemies, or between territories which functions as a limit. A frontier implies expansion, a horizon line that serves as a beacon rather than an end point.

Past empires established their dominance in an environment where resources were thought to be finite, extending their supremacy to capture as much of the world's wealth as possible, to the detriment of their rivals. Now, though, the United States made a credible claim to be a different sort of global power, presiding over a world economy premised on endless growth.[15]

The modern American frontier is less characterized by hoarding resources and more so as a self-appointed keeper of the myth of endless economic growth on a global scale, facilitated by natural and financial resources within the continental, neo-colonial, and now atmospheric boundaries. "To talk about the frontier is also to talk about capitalism, about its power and possibility and its promise of boundlessness."[16] The space exploration enthusiasts of the middle of the twentieth century already understood the relationship between frontiers and capitalism, and that to protect the internal mechanisms of capitalism from reaching its planetary limits, one must tap into the exploitable resources in the stars. The movement to extend the national economic reach skyward is not limited to profit-seeking, but rather the economic opportunity supports the effort to bring human life closer to the frontier of interplanetary habitation. The border between Earth and other worlds is thus transformed into a frontier, a cold, black horizon that need only be conquered through technological innovation and cowboy daring.

Valerie Olson's work in *Into the Extreme* argues there is a project of redefining what is or is not included in the concept of the planetary. Her work provides an anthropological analysis of space scientists in the United States.[17] This definition is important because it provides a cognitive map which allows scientists to conceive of boundaries in specific ways. In defining NEOs as part of the planetary, she argues that people tend to focus on two different factors: (1) the realization of increasing risk from NEOs on the Earth's ecosystems and (2) the realization that NEOs can be an ample source of limited resources used either for Earth-based production or as a cheaper means of providing resources for human settlements in space. The first defines the planetary using the logic of ecology and risk assessment. Since meteors can be a catastrophic force, as was seen in the Helyabinsk meteor event in Siberia in 2013, NEOs are already part of the broader web of planetary life. The presence of a threat allows for the development of planetary administrative systems that, following the work of Timothy Luke, allows for the management of meteors and other NEOs through a lens of eco-governmentality administered through the neoliberal military security state.[18] To Luke, eco-governmentality is the use of administrative logic to manage and regulate the natural world for the benefit of the state and the capitalist economic system. The desire to manage the ecosystem inevitably leads to the management and regulation of the biopower of human and nonhuman life. The use of aeronautics and military contractors to develop outer space resource extraction technologies brings together climate change, corporate property grabs, and national security as part of an imperative to "protect" the American people from a potentially dangerous situation. The use of fear and emergency powers follows the internal logic of what Naomi Klein calls "disaster capitalism."[19] In this case, pre-disaster capitalism is the new investment opportunity, with venture capital sinking significant resources into privatizing and neoliberalizing outer space as responses to climate change and under the guise of protecting the world from the threat of a meteor strike. The Cold War anxieties about militarizing space are refashioned as resource grabs and securing outer space resources, not just using outer space as a venue for satellite and weapons posturing.

The second justification, which requires the future of humanity to be a multi-planetary, if not galactic, takes its cues from the "frontier myth." In this myth, first described by Frederick Jackson Turner, the maintenance of democracy and American capitalism required an ever-expanding frontier. This frontier served as a pressure valve for the country's social and class conflicts by pushing unemployed white workers to the newly colonized western lands. It was also an essential aspect of constructing and defining the "American spirit" of seeing citizenship born in a baptism of struggle, hardship, and conquest. Exploring the edges of American frontiers would "reduce racism to a remnant and leave it behind as residue. It would dilute other social problems as well, including poverty, inequality, and extremism, teaching diverse people how to live together in peace."[20] While the physical frontier of North America was closed after the completion of the colonial process as Manifest Destiny, the limits of the frontier were overcome via the development of the United States as a multi-continent colonial and imperial force. While space as the "final frontier" has been in the American psyche since the beginning of the space race in the 1960s, that earlier struggle was largely about national glory and military superiority in the context of the Cold War. With the collapse of the Soviet Union and the unfettered growth of neoliberal capitalism, the values of glory and heroism have, as Marx would put it, "melted into air" leaving behind only cold economic calculations and profit margins. With the development of such mindsets, the ability to redefine outer space is not only a political project, but also an economic one. In fact, space race enthusiasts have used frontier language for decades.

In *Space Colonies*, Stewart Brand makes a case for using the term "colonization" to describe the new movement to send humans to space because that kind of frontier motif seemed the most apt.[21] Rick Tumlinson, co-founder of several space-related organizations, such as Deep Space Industries and the Space Frontier Institute, hosted the Space Cowboy Ball in 2019. The ball honored Jeff Bezos, CEO of Amazon and space technology company, Blue Origin, with the Space Cowboy Award, which "recognizes those who have ridden out ahead of their field and made significant contributions to the opening of the frontier."[22] Given the language of contemporary space funders, it is not a stretch to trace a line between the ideological drivers of early American frontier colonialism and those of contemporary space exploration. The neoliberalization of space comes with a complicated relationship to borders. While space funders see their projects as an opportunity to transcend planetary boundaries and send neoliberal economic and political systems into space beyond the boundary of Earth's atmosphere, they depend on a legal environment that protects their investment through national boundaries based on property rights. The frontier capitalism of the new space race meets the legal framework of establishing borders between property holders. Who on Earth gets to own, lease, or otherwise access and profit from the outer space frontier?

Carl Schmitt in *The Concept of the Political* provides a philosophic justification to construct boundaries and borders. Schmitt discussed the need to have a well-defined "enemy" and "friend," which to him is the foundation of "the political," but the core of the concept works to provide insight into other boundary settings.[23] To Schmitt, for a political category to have value there needs to be a mechanism of exclusion and a clearly defined line. For instance, if "citizenship" is to have value there needs to be a "non-citizen" who is excluded from the benefits of citizenship. Without that contrast, there is no need to speak of "citizenship" because it is not a category that has any value.[24] Schmitt's infamous friend/enemy dichotomy meets the openness of the frontier myth built into the justification for exploring NEO mining and space colonization. While the narrative of exploration provides a picture of resource availability to stimulate exploration and energy production to satisfy planetary needs, the treaties and legal landscape of public and private investment in these same activities reinforce Schmitt's boundary between the friend and enemy, or between producing to support one's nation to strengthen it against one's enemies. The United States as a self-proclaimed "superpower" requires constant expansion – both ideological and territorially – to sustain its position. As Wolin defines it,

"Implicit in that declaration ["Superpower"] is a reformulation of the nation's identity: it stands for sheer power, economic and military, that is measured by a global standard rather than the nation's constitution; freed not only from constitutional democracy but from any truly political character."[25]

Wolin contends that western expansion required "a technology of power that scan make occupation and rule effective."[26] Western expansion is now pushed up to the stars and not just toward foreign shores. Although the United States and the associated corporations supporting outer space exploration and mining claim these efforts are for the global good, the desire to go to space suggests national economic and military gain, rather than another version of the global unity attempted in the aftermath of the world wars.

In *Land and Sea* Schmitt provides a neo-conservative foundation for international relationships through an engagement with historical struggles between land and sea empires. He opens this work by defining the human as a land-based animal. In Berman's interpretation, the human are people with roots in the soil, rather than the rootless people who cannot claim deep ties to the land and the boundaries thereof. Given Schmitt's affiliation with the Nazi party, it should come as little surprise that the landless people, the sea-people, are the "wandering Jews" and the British Empire.[27] The land is under the dominion of a particular people and nation, while "the high sea is free, i.e., state-free and subject to the authority of no state dominion."[28] Schmitt goes on to elaborate his theory of land versus sea empires. This is, according to Fluss and Frim, a metaphorical engagement with Hobbes, via monsters. They write:

> These beasts are a pair of opposites: Behemoth is autochthonous, representing the stable order of earth-bound peoples. Leviathan is thalassocratic, embodying the fluid dynamism of seafaring peoples. Behemoth signifies terrestrial empires, while Leviathan suggests commercial trade and exploration. The former stands for traditional, divinely sanctioned state authority, the latter for the spirit of pirate-capitalist enterprise (what Schmitt calls 'corsair capitalism').[29]

This monstrous battle between different forms of empires, to Schmitt, is central to understanding the struggle to understand and define the "nomos of the earth." He writes that advances in politics and empires are the result of "new lands and new seas into the visual field of human awareness." To make this point even stronger he writes that "...all important changes in history more often than not imply a new perception of space." Industrialization, especially in England, a sea empire, radically changed the country. The sea and land powers changed again with the advent of radio and airplanes, which introduced a third dimension, or the first planetary spatial revolution.[30] Movement into the air includes movement into outer space, and global communication through airborne radio waves encircling the Earth. This new nomos of the Earth, partially fueled by fire through engines and rocketry, presents a fundamental shift in the planetary powers. As Berman notes in his introduction to the volume, whatever else Schmitt is doing, he provides "a planetary materialism that grounds culture in the elements of the environment. *Land and Sea* is about nothing if not the human dependence on the natural world in which we live."[31] These technological advancements open up empire and conflict beyond the boundaries of land and sea, and the fundamental connections between people, resources, and their spatial relations with the planet. In a Schmittian sense, the re-imagination of the planetary constructs a new conception of space, and, in doing so, it opens up radical space for expansion, political maneuvering, and potential conflicts. Schmitt described the United States as the new sea-based empire with which the world would contend, and one that would supplant England as the sea-based global empire. Practically, the 2015 Space Act opens up space exploration to US corporations and imposes on NEOs the property law of the United

States. The act, which requires the US government to protect the economic interests of US space exploration corporations, is a continuation of the long line of neoliberal policies in which the State uses its power in order to protect the economic interests of corporations. While the OST is meant to reinforce international cooperation and "benefit for all mankind" regardless of a nation's ability to go into space, the practice of doing so is dependent upon neoliberal logics of open markets and the flow of capital and resources that inevitably benefits very few at the publicly funded expense of many. "Scientific and technological developments, depending on their content, could eventually flow to all nations by way of sale through global markets."[32] The "rising tide" argument depends upon non-state entities cooperating with states to only conduct "peaceful" and economic actions on asteroids and other celestial bodies, which assume that non-state actors are not a proxy for a sovereign state. Somehow, private companies are both supervised by their home nations and permitted to operate independently as though they are sovereigns as well. The Space Act nudges the boundaries of the OST by leaving open the question of the place of non-state actors. The OST does not foreclose private sector participation; public-private partnerships are increasingly used to contract out work that NASA can no longer afford. Some of these private sector partners are responsible for design and componentry, while others, like Bezos's Blue Origin and Musk's SpaceX, have more ambitious plans to bring people and economic opportunity to outer space themselves. The rise of public-private partnerships in the new space race will force a reevaluation of the parameters of the OST, and what counts as violations of the prohibition against territorial claims and resource use that is not for the good of all humanity. Current treaties to share international waters and Antarctica are cited as examples of workable solutions to manage space as a common pool resource, which further reifies outer space as a resource that can be managed and extracted. Saletta and Orrman-Rossiter even propose repurposing the licensing and royalties system of permitting private sector extractive activities on public lands as a model, although they are yet unclear how to ensure that royalties collected will be fairly distributed to everyone with a stake in outer space, which quite literally includes everyone.[33] What is clear, however, is that public space programs are increasingly privatized, despite claims that asteroid mining and space exploration are demonstrably good for the global public, and the dependence on the private sector to define and manage public needs exemplifies the shift toward neoliberalizing outer space. Following Schmitt, this process does not function to create a global "friend" but to instead maintain nationalist boundaries for the United States, which are essential contrasting US identity and property claims from those of other nations, and economic imperialism directed toward the heavens.

Neoliberal space policy

In the case of NEAs, the new perception of space is actually a redefining of outer space as inner space. This redefinition is a shift in legal perspective of space. In expanding the category of space, the 2015 Commercial Space Launch Competitiveness Act ("Space Act") undoes a half century of international space law. The original realm of space law was constructed during the Cold War and, like the limits on nuclear weapons, was an attempt to limit the possibility of "hot" conflicts between the two hegemonic forces. The outer space treaties – the first of which was signed by the United States, the United Kingdom, and the Soviet Union in 1967 – were primarily about limiting nuclear weapon proliferation into space, but they also effectively non-commodified space by requiring all space exploration of foreign bodies and planets to be in the shared interest and benefit of the entire planet. This limited the ability for corporations to privatize outer space, and limited any states' power to manage, regulate, or own celestial entities. The Moon Agreement, which "prohibits private ownership over lunar natural resources," has yet to be ratified by the United States.[34] According to legal reviews, the OST and the Space Act are at odds with one another. The OST prohibits "national appropriation" of celestial bodies, including use by private entities, and the Space Act asserts that private use does not amount to sovereign claims on these bodies. "In other words, Congress effectively interpreted the OST to prohibit only national appropriation by claim of sovereignty and by use or occupation, an interpretation it understands to be in conformity with international law."[35] In one interpretation of these space treaties, "...unlike the Moon Treaty, however, the OST does not explicitly prohibit commercial development," which the author claims can account for the willingness of major countries to agree to the OST but not the Moon Treaty.[36] Another challenge to the application of these space treaties is the decline in national agencies as the primary actors in space exploration, and the rise in private sector investment. Space exploration as a "good for all mankind" is increasingly an open frontier for the private sector to establish the infrastructure of a space market. In a 2009 NASA report, "Review of Human Spaceflight Plans Committee – Final Report," the committee suggests that the way to bring down the cost of human spaceflight programs is to "turn this transportation service to the commercial sector" and to "establish a new competition for this service, in which both large and small companies could participate."[37] The committee suggests that NASA could focus its resources on developing new technologies and leave the work of transport to the private sector.

The 2015 Space Act was passed in an attempt to change the legal property rights regime of Near Earth Objects, effectively, wrapping them within the legal framework of the United States. While neither as collaborative nor as inhumane as the 1884 Berlin Conference, where the great European powers split the African continent among each other, the current law follows in its footsteps by claiming large tracts of space and its resources as the colonial ownership of mostly US corporations. Unlike in previous colonial distributions of land, this project does not follow the mercantilism of the late nineteenth century but rather the neoliberal politics of Reagan and Thatcher. Neoliberalism, as understood by Wendy Brown, is not about shrinking the state, nor about allowing free market logics to govern the state but is instead an active process of state reformation. The neoliberal state actively works to both construct and monitor marketized spaces that create and attempt to reconfigure subjectivity toward market logic. Brown in *Undoing the Demos* contends that the neoliberal revolution is largely about the active construction of free markets as neoliberalism and, compared to the more classical liberalism of Adam Smith, does not believe that free markets are natural. Like all social relationships, the features of neoliberalism are formed and forged through social pressures and state power. The state's role is not to just step back and allow the market to govern but to constantly create and recreate the needed market relationships.

When it comes to outer space, this means not only the enclosure of land and resources into private hands but also turning over the planning and imagining to the private sector. In David Valentine's anthropological study of the "NewSpace" advocates – those entrepreneurs, engineers, and enthusiasts who fund and produce private space companies and technologies – he finds that these advocates:

> Urgently insist that entrepreneurial human settlements in outer space will resolve these problems [environmental degradation] by enabling clean power through space-based solar power generation, the end of resource wars through asteroid mining, and growing human prosperity by the expansion of free markets into space.[38]

Valentine argues that those agitating for the commercialization of space are not simply advancing a plan to extend capitalism to the stars; they are legitimately concerned with climate change, biodiversity loss, and making human consciousness multi-planetary. Valentine makes a compelling argument to consider the private sector space plans in its fullness as a site of political ideology, cosmology, and economics. In the description of space summits and conferences, where entrepreneurs pitch their space-centric plans and the attendant business details, it becomes clear that there is no space industry without finance capital. The two are drawn together in a kind of purposeful inevitability; that so long as outer space planning is developed in a neoliberal state, the plans themselves will bear the mark of this kind of short-term, profit-driven planning, even if there are other dreams attached. Valentine argues that "we should engage with those [cosmological] visions in *their own terms*, and not foreclose them within the unfolding of a story that we already know (the eternal success of neoliberalism or the inevitability of an environmental and socialist revolution)" which requires "a willingness to accept NewSpace as a serious social movement whose plans and activities may produce at least some of the outcomes it claims."[39] Valentine is not wrong to ask that the plans for a space-faring civilization be taken seriously. However, at present, they are largely inseparable from the neoliberal political and economic context in which these plans are hatched. To acknowledge this context does not reduce these ideas to merely a veneer over the real project of extending capital into space. Instead, we must understand the projected "space futures," from Mars colonies to asteroid mining, are currently imagined within an economic and governing system that precludes the more creative and pressing work of linking space-faring dreams to the crises of climate change precipitated by late capitalism and neoliberalism.

The core value of the 2015 "Commercial Space Launch Competitiveness Act" is: "To facilitate a pro-growth environment for the developing commercial space industry by encouraging private sector investment and creating more stable and predictable regulatory conditions, and for other purposes." The focus on private sector investment and the use of state power both to create a market and to stabilize and protect that market are clear examples of neoliberal policy. The state uses "public safety" as part of a broader project to deregulate an industry and divert government funds toward private corporations. In the case of NEAs, the threat of potential meteors was used as a precursor for the policy. The bill asserts that it is designed to "protect public health and safety, safety of property, national security interests, and foreign policy interests of the United States." Much like the expansion of the private security and intelligence apparatus following September 11 and the invasion of Iraq, the US government used the threat of public harm to justify transferring public funds to private contractors. Unlike many neoliberal policies, which seem to be neutral on the nationality of private corporations, neoliberal space policies tend to embrace a certain form of nationalist neoliberalism, a type of deregulation that works to both deregulate national capital while strengthening the state's power to exclude and, at least symbolically, strengthen claims around national borders. The 2015 bill does this by only offering the regulatory protections for US corporations to "facilitate commercial exploration for and commercial recovery of space resources by United States citizens." This is in line with Russian debates and discussions around Roscosmos and the United Rocket and Space Corporation. In the Russian example, the state created government-invested private corporations which were given state resources but were allowed more legal and policy leeway. Of course, such a program did not work well and in 2013 there was a push to renationalize the corporation amid rampant corruption and wasteful spending. Russia, which is well known for its oligarchic corruption, formalized the flow of public money to private hands. The US policy, unlike the Russian, is not focused on a singular corporation but instead on developing a broad corporate partnership with the US government – primarily for wealthy Silicon Valley executives and military technology and weapons contractors. The law states that it is going to "discourage government barriers to the development in the United States of economically viable, safe, and stable industries for commercial exploration for and commercial recovery of space resources in manners consistent with the international obligations of the United States." Bezos and Musk agitate for the same, and Silicon Valley start-ups and venture capital firms are known to support the public-private partnerships that transformed Silicon Valley from orchards to a global tech center. While the law formalizes the neoliberal policies as an aspect of planetary defense against meteors, in practice the law is primarily about resource extraction:

> A United States citizen engaged in commercial recovery of an asteroid resource or a space resource under this chapter shall be entitled to any asteroid resource or space resource obtained, including to possess, own, transport, use, and sell the asteroid resource or space resource obtained in accordance with applicable law, including the international obligations of the United States.[40]

This language tries to walk a line between the extracted resource and the asteroid or celestial body from which it is extracted. Even if it were possible to separate the two, legally, conceptually, and physically, this leaves space for the next generation of mining companies to decimate outer space resources without taking responsibility for the environment that provides these economically valuable resources. Separating NEOs and NEAs into industrial metabolic ingredients only extends the problematic, neoliberal environmental use that plays a role in our current environmental crises on Earth.

Conclusion

This chapter argues that the public and private sectors are taking seriously the opportunity to expand space exploration to include asteroid mining and space colonization, and as such these endeavors require critical attention to the questions and answers that are getting produced within a neoliberal framework for human/environmental relations. It does not set out to argue for or against space exploration, or even asteroid mining. Rather, it seeks to create a critical framework for analyzing such endeavors so that they are sufficiently politicized and considered an important part of the broader, social inquiry into surviving climate change and becoming a multi-planetary species. The framework includes questions of boundaries (earth and outer space; the limits of sovereign national claims in outer space) and a re-reading of outer space treaties and public laws that seek to govern economically viable claims to space resources that do not imply that neoliberalism is the only possible system in which these decisions should be made. To paraphrase and play with Jameson's claim that it is easier to imagine the end of the world than the end of capitalism, this chapter posits that it is easier to imagine lassoing an asteroid and supporting human colonies on Mars than it is to imagine addressing the dangers of climate change, and the potential of space exploration, from an economic and political system that rejects the values of neoliberalism. Perhaps the only thing scarier than leaving the capsule to explore the cosmos is leaving the proverbial capsule of a neoliberal status quo.

Notes

1 "NEO basics" https://cneos.jpl.nasa.gov/about/target_earth.html.
2 "NEAs as resources" https://cneos.jpl.nasa.gov/about/nea_resource.html.
3 "Asteroids: Overview" https://solarsystem.nasa.gov/asteroids-comets-and-meteors/asteroids/overview/?page=0&per_page=40&order=name+asc& search=&condition_1=101%3Aparent_id&condition_2=asteroid%3 Abody_type%3Ailike.

4 Reed Elizabeth Loder, "Asteroid Mining: Ecological Jurisprudence Beyond Earth," *Virginia Environmental Law Journal* 36, no. 1 (2018): 276–317.
5 James O'Connor, *Natural Causes: Essays in Ecological Marxism* (New York, NY: Guilford Press, 1998), 165.
6 O'Connor, 165.
7 O'Connor, 164–65.
8 Aaron Bastani, *Fully Automated Luxury Space Communism: A Manifesto* (New York, NY: Verso, 2019).
9 Aaron Bastani, "The World Is a Mess. We Need Fully Automated Luxury Communism," *New York Times*, June 11th, 2019.
10 John Bellamy Foster, "Marx, Value, and Nature," Monthly Review, https://monthlyreview.org/2018/07/01/marx-value-and-nature/.
11 Steven Vogel, *Thinking Like a Mall* (Cambridge, MA: The MIT Press, 2015).
12 Herbert Marcuse, *One-Dimensional Man* (Boston, MA: Beacon Press, 1991).
13 Kathryn Yusoff, *A Billion Black Anthropocenes or None* (Minneapolis: University of Minnesota Press, 2019).
14 Greg Grandin, *The End of the Myth* (New York, NY: Metropolitan Books, 2019), 2.
15 Grandin, 3–4.
16 Grandin, 8.
17 Valerie Olson, *Into the Extreme* (Minneapolis: University of Minnesota Press, 2018).
18 Timothy W. Luke, "On Environmentality: Geo-Power and Eco-Knowledge in the Discourses of Contemporary Environmentalism," Cultural Critique 31, (1995): 57–81. doi:10.2307/1354445.
19 Naomi Klein, *The Shock Doctrine* (New York, NY: Picador USA, 2007).
20 Grandin, *The End of the Myth*.
21 Stewart Brand, *Space Colonies* (New York, NY: Penguin, 1977).
22 "Spacefund Founder," https://spacefund.com/spacefund-founder-presents-jeff-bezos-with-space-cowboy-award/
23 Carl Schmitt, *Concept of the Political* (Chicago, IL: University of Chicago Press, 2007).
24 Of course, this does not mean that borders and binaries are essential as many radical politics – such as the history of anarchism and Marxism – are about abolition categories either by making a concept impossible (such as the way that communism would abolish class distinction) or by destroying the disparate value (such as anti-racist projects around abolishing whiteness and white supremacy).
25 Sheldon S. Wolin, *Democracy Incorporated.* (Princeton, NJ: Princeton University Press, 2008), 60–61.
26 Wolin, 61.
27 Carl Schmitt, *Land and Sea* (New York, NY: Telos Publishing, 2015).
28 Samuel Garrett Zeitlin in Schmitt, lvii.
29 Harrison Fluss and Landon Frim. "Behemoth and Leviathan: The Fascist Bestiary of the Alt-Right." Salvage, http://salvage.zone/in-print/behemoth-and-leviathan-the-fascist-bestiary-of-the-alt-right/?fbclid=IwAR20VxZU8OAsinxAV-eqPULxJcMYUG-yn-aDcfVGslzHJKna6dfqoYd4CSY.

30 Schmitt, *Land and Sea.*
31 Zeitlin in Schmitt, xxviii.
32 Jack Heise, "Space, the Final Frontier of Enterprise: Incentivizing Asteroid Mining Under a Revised International Framework," *Michigan Journal of International Law* 40, no. 1 (2018): 8.
33 Sterling Morgan Saletta and Kevin Orrman-Rossiter, "Can Space Mining Benefit All of Humanity?: The Resource Fund and Citizen's Dividend Model of Alaska, the 'Last Frontier,'" *Space Policy* 43 (2018): 1–6.
34 Virginie Blanchette-Séguin, "Reaching for the Moon: Mining in Outer Space," *NYU Journal of International Law and Politics* 49 (2017): 959.
35 Heise, "Space, the Final Frontier of Enterprise."
36 Loder, "Asteroid Mining."
37 "Review of Human Spaceflight Plans Committee – Final Report," www.nasa.gov/pdf/396093main_HSF_Cmte_FinalReport.pdf.
38 David Valentine, "Exit Strategy: Profit, Cosmology, and the Future of Humans in Space," *Anthropological Quarterly* 85, no.4 (Fall 2012): 1045–1067.
39 Valentine, 1065.
40 *Space Resource Commercial Exploration and Utilization, U.S. Code 51*, Ch. 513. https://uscode.house.gov/view.xhtml?path=/prelim@title51/subtitle5/chapter513&edition=prelim.

5 Conclusion
Robert E. Kirsch and Emily Ray

The contributions in this volume very clearly do not fit in with a standard environmental policy analysis framework. While that research is of course valuable, the contributions to this volume infuse a critical political and social theory to show the scope of the challenges for a world undergoing climate change and where many environmental political theorists are making contributions. All of the chapters, in their own particular way, make the case that business as usual cannot yield a better world, and that waiting for a technological fix that reverses environmental degradation is not a worthwhile use of time and is bound to end badly. Concluding this volume will be an eye toward the current context, the role of critical theory, and charting out some of the directions Environmental Political Theory (EPT) can go or is already going.

The role of critical theory

All of the chapters in this volume brought in a number of critical political and social theorists. Hopefully, a convincing analysis obviates the need to justify doing so. It is not, in other words, helpful to cordon "environmental theory" from other types of theory. Critical theory is interdisciplinary and, at its best, engages in an immanent critique of society that shows the inadequacy of a regime to its stated goals or accomplishments.[1] Because it lacks traditional disciplinary boundaries and takes social and technological relations as they are (rather than idealize them), critical theory can help make an incisive diagnosis of the current context. This of course does not mean that anything goes, and all social theorists are applicable to all environmental concerns, but it does mean that building a careful case of integrating critical theory into environmental analysis keeps these strands in conversation with each other, and that can yield some novel insights. Many of the issues of political theory still resonate in a variety of social registers, not least of which are environmental. As the contributions in this volume have shown, theoretical concepts such as sovereignty, borders, deterrence, technics, and consumption, among others retain their currency when addressing the enormous social complexity of responding to energy extraction, production, and consumption. In fact, reducing environmental problems to a narrow lens of technological challenges is just the kind of ecomodernism that is critiqued by Luke; critical theory expands the horizon of analysis to better address that complexity. This is to say nothing of the fact that many critical theorists themselves wrote extensively about ecological concerns, because of the obvious relationship between ecology and society.

With that in mind, this volume is only a contribution to this ongoing critical conversation and is in no way exhaustive or final. As this field that can loosely be called a political economy of energy (or perhaps a political economy of extreme energy production) continues to develop, it is easy to imagine drawing in Foucauldian concepts of ecogovernmentality, Gramscian notions of counter-hegemonic practices, or Veblenian analyses of wastefully conspicuous consumption, among many others. The goal here is not to give a final list of the proper critical theorists who should be consulted but to implore scholars to continue integrating critical theory into environmental analysis.

One of the significant contributions of critical theory is the framework for challenging the premises of ecomodernism and ecopragmatism. The chapters in this volume take aim at the promise of technological fixes and steady progress through market mechanisms and green consumption. As Marcuse argues, the destructive tendencies of technologically advanced industrial society produce short-term comforts at great ecological expense.[2] Ecopragmatism advocates for increased control over consumption and production to account for the expense of lost "ecosystem services" in the course of regular economic activity. This approach to sustainability is predicated on mastery over nature and anticipates the unpopularity of promoting a radical shift in the basis of society altogether. None of the chapters in this volume suggest a fantastical human/nature relationship perfectly in balance and without a trace of exploitation. In fact, these chapters are not particularly prescriptive but rather take up the spirit of critical theory by stripping away the stories of endless progress toward a liberal democratic world fully under the control of the Western empires and instead reveal the inner workings of Western liberal democracies to find the destructive forces animating them. Ray and Parson do not claim that off-earth mining is unthinkable but rather that the motivation and practices to mine near-earth objects speak to the neoliberal approach to problem-solving climate change while maintaining current economic and social relations. Similarly, Kirsch's paper points to the civilizational roots of ecological domination, which provides a basis for analysis of contemporary wasteful and class-stratified use of energy. Luke's paper demonstrates that ecopragmatism is more than ineffective in the face of climate change, but it has promoted the destructive activities that have worsened the crisis while mitigating the worst effects for a large enough number of people to preserve the system of production. Existing policies and policy proposals to address the many faces of climate change fall woefully short of the large-scale changes necessary to maintain a planet habitable for many current species, humans included. The volume pushes beyond handwringing over the climate crisis and offers theoretical tools for addressing the many tasks at hand, including confrontations with, and proposals for, public policy.

Current policy initiatives

As shown above, a critical perspective can be a discerning tool to judge the effectiveness, desirability, or impacts of certain environmental policy prescriptions from within their larger social contexts. With that in mind, this kind of theorizing can be used to understand, shape, or critique policy proposals as they take shape or are implemented. While not directly taken up by the chapters in this volume, the theoretical insights can help shape the terrain of the debate of a suite of policy imaginaries that can be considered as part of a "Green New Deal" as well as its alternatives. From Ray and Parson, it is easy to conceive of a neoliberal, market-based solution to climate change where everybody takes responsibility for their own well-being. With such a regime in place, there is no need for a Green New Deal. Rather, this ecolibertarian preference would valorize individualized responses to climate change like doomsday prepping, media spectacle consumption-based approaches, and population management. There is also an elite version of this consumption-based approach that includes, much like the rockets of Musk and Bezos, building "offworlds" on other planets, seasteading on stateless oceans, or manufacturing islands as a respite from the chaos. These individualistic responses to climate change may appeal radical – rockets and private islands to ride out climate chaos – but they are an extension of the same techno-optimistic ecomodernist approach to politics that assumes carbon markets and electric cars will herald a new world adapted for climate change, even if only for the wealthy few. Ray and Parson convincingly argue that even though the ruling class wishes to break free from the rule of the state, its absence is managed. That is, even in outer space, claims of sovereignty and matters of territories and boundaries all come with the private rockets. It seems unlikely that an adequate response to the problems of climate change is to simply let the rich do whatever they like, and a critical perspective rather quickly shows the inadequacy of such an approach.

At the same time, advancing a Green New Deal simply because it is preferable to an ecolibertarian hands-off approach does not mean that all policy suggestions are by definition desirable. A Green New Deal would necessarily exceed domestic policy and impact international relations as well, including trade and technology sharing, but should not be dependent on assuming enough research and development will yield the technology to "decouple" resource extraction and the energy required to sustain lifestyles. As the past decades of United Nations climate change conferences have demonstrated, attempts at accountability and global climate justice between industrialized and developing nations are a serious feature and sticking point of environmental policy making, to say nothing of the inability to meet the inadequate metrics set by, for instance, the Paris Agreement. Therefore, a Green New Deal with global implications, constructed by the same people with literal and figurative investments in the political and economic systems as they exist, will have enormous challenges living up to the calls for radical change. It is indeed easy to conceptualize an ecopragmatist Green New Deal with its technological rationality underpinnings that, as Luke emphasizes, is only a politics of delay and delusion. What, then, might a non-deluded Green New Deal look like? The necessity of system-wide change comes to the fore in all of the contributions. It is not enough to put faith in the technocrats to fix the environment through "wise use," nor to place faith in the earthly gods of Silicon Valley to engineer a quick fix. Neither, still, is a mass mobilization of military might. Whatever the specific policies end up being debated in environmental legislation (Green New Deal, neoliberal ecomodernism, or militarization), this critical approach centers the policy debate on its democratic foundations. Namely, how can a Green New Deal change social relations of production and take into account the broader lived experience of citizens in terms of health care and environmental racism, fair housing opportunities and tending to the urban/rural split, as well as reimagining public utilities and employment.

Directions for EPT

EPT is an umbrella for different approaches to describing, explaining, and critiquing the human/nature relationship and the political, economic, and social stakes of those relations. Common to EPT writing is critical attention to power structures and dynamics, especially as they intersect with global environmental issues, including interspecies relations and rights, environmental justice, and ecosystem management. EPT scholarship is not exclusively critical, and sometimes suggests ways forward for humans to live on the planet less destructively, including toward each other.[3] The chapters here push EPT themes to the edge of energy extraction and production, and, in doing so, center social and political theory as the basis for critique and analysis. Since the 1980s, EPT has been advancing ecological concern beyond ecomodernism and policy prescriptions and has made a case for foregrounding environmental issues as worthy of serious inquiry and analysis through various political theoretical lenses. This volume puts forward an interdisciplinary theoretical study of the political economy of energy that does not rely on traditional economics or policy analysis to contain energy production and consumption within the standard of cost/benefit analyses, or as pieces to fit into the puzzle of sustainable capitalism in a neoliberal democracy. Luke has made significant contributions to EPT, including his works on environmentality, or the rationalities and technologies for global ecosystem management. His work in this volume builds on the foundation of Foucault and the Frankfurt School to contextualize environmental crises as symptoms of the scientific rationalization of life, disciplining of subjects, and the one-dimensional world of politics, economics, and intellectual life that makes ecomodernism appear as the only avenue for managing the environment. Ray and Parson pick up the critique of rationalizing the environment as a resource pool in their exploration of off-earth energy and national sovereignty as extraterrestrial economic domination. They push EPT to consider the logical extension of global ecosystem management into the stars and reach back into public policy – relevant to but not the central focus of EPT – to demonstrate that the global superpowers have already been concerning themselves with off-earth resource management, and it's time environmental scholarship did too. In the spirit of interdisciplinarity, Kirsch brings together Mumford and Bataille to generate a basis for a political economy of energy. Instead of emphasizing fossil fuel production and consumption, he likewise looks to the stars, or rather one big star, as the source of energy we on earth cannot help but waste. He turns to Bataille's theory of energy wasting as a new avenue for addressing the structural problems with energy management. Taken together, these chapters build on a critical and timely platform of EPT scholarship, and offer new theoretical frameworks for analyzing our political, social, economic, and ecological relations with energy. In the spirit of EPT's bold pursuit of greening political theory, this volume pushes the field to reexamine the legacy of ecomodernism, to account for the next generation of this thinking in the form of ecolibertarianism, and to include outer space in the sphere of political environmental concern.

Notes

1 Max Horkheimer, *Critical Theory: Selected Essays*, trans. Matthew J. O'Connell (New York, NY: Continuum Publishing Corporation, 1975).
2 Herbert Marcuse, *One-Dimensional Man* (Boston, MA: Beacon, 1968).
3 Teena Gabrielson, Cheryl Hall, John M. Meyer, and David Schlosberg, eds., *The Oxford Handbook of Environmental Political Theory* (Oxford, UK: Oxford University Press, 2016).

Index

Abbey, Edward 29
abolitionist movement 42–3
activism 57
air pollution 36
alienation: of humanity 16; Marx's theory of 15; overcoming 16
American frontier colonialism 63
Archeology of Knowledge (Foucault) 60
asteroids 55–7; mining 2, 57, 58

Bastani, Aaron 57
Bataille, Georges: "fulfillment of things" 16; general economy 5, 12, 15, 21; "general economy" theory 3; sovereign 18
Baudrillard, Jean 46, 47
Behemoth 65
Berlin Conference, 1884 68
Berman, Russell 65
Bezos, Jeff 2, 3, 63, 70
A Billion Black Anthropocenes or None (Yusoff) 60
Blue Origin 66
Bookchin, Murray 29
Brand, Stewart 2, 3, 25, 28, 31, 34–5, 37, 38, 48, 63; *The Whole Earth Catalog* 30, 31; *Whole Earth Discipline* 29, 30
Braungart, Michael 32
Breakthrough Institute 25, 28, 38–41, 44, 46–8
Brown, Wendy 68

capitalism 58, 60
carbon dioxide (CO_2) emissions 49
Carson, Rachel 28, 29
celestial bodies 57
citizenship 64
Clark, Brett 55, 58
clean energy technologies 37
climate change 25, 57, 74; activism 26; challenges 26–7; movements 30
colonization 63
Commercial Space Launch Competitiveness Act 56, 69
Commoner, Barry 29, 42
The Concept of the Political (Schmitt) 63
consumption: good and bad duality of 19–20; nonproductive 17; political economy of 5–21; productive 3, 15–17; sovereign 12–18; surplus 12–18
contemporary capitalism 59
corporate sustainability 27
cradle-to-cradle (c2c) design 32, 33
critical theory role 74–6
crypto-abolitionism 43

"The Death of Environmentalism: Global Warming Politics in a Post-Environmental World" (Schellenberger and Nordhaus) 27
decarbonization policies 26
decoupling 35, 36, 41–5
Deng Xiaoping: "Four Modernizations" 29

Index

disaster capitalism 62
Discipline and Punish (Foucault) 60
Drake, Edwin 42

earthly gods 2–3, 5
eco-governmentality 75
ecolibertarianism 79
ecological crisis 59
Ecological Rift (Foster, York, and Clark) 55
ecomodernism 1, 2, 25, 27–30, 45, 75
Ecomodernist Manifesto 33
ecomodernists 27, 29, 32, 33, 39, 43, 49
ecopragmatism 1, 2, 25–30, 75
ecopragmatists 27, 30–1, 40
Edison, Thomas 42
Ehrlich, Paul 29
energy 7–9; Bataille's theory of 19; consumption (*see* consumption); hydrocarbon 31; nuclear 29, 31, 42, 43; solar 12–13, 43
energy extractivism 25
energy slaves 42
Environmental Grantmakers Association (EGA) 27, 28
environmental movement 57
environmental policy: current mode of 3–4
environmental political theory (EPT) 1, 74; directions for 78–9; scholarship 78
environmental politics: climate change challenges and 25–6
Environmental Progress 29, 36, 47
extraction: activities, private sector 66; capitalism 57; destructive 17; industries 55; political economy of 13, 15, 18; sites 13
extraterrestrial extraction 3, 20
extraterrestrial private mining 55
extraterrestrial resource extraction 55

Fluss, Harrison 65
fossil energy 26, 29, 42
Foster, John Bellamy 55, 58
Foucault 60
"Four Modernizations" 29
Friedman, Thomas 45, 46
Frim, Landon 65

frontier myth 62
Fukushima Daiichi nuclear disaster 38
Fuller, Buckminster 38, 42

garden megacities 39
general economy 3, 5, 12, 15, 21
geoengineering 20, 31
global warming 25, 26
"gloom and doom environmentalism" 32
GMO-based agriculture 47
"good Anthropocene" 26, 33, 34, 38, 39, 48
Goodman, Paul 38
Grandin, Greg 61
Great Financial Crisis of 2008 20
Great Recession 38, 46
green business 27
Green New Deal 45, 76, 77
green nuclearity 39
green pragmatists 37
growth 13, 14

"Half-Earth" (Wilson) 37–8, 48
Hansen, James 26, 27, 29, 49
Haraway, Donna 60
Hardin, Garrett 29
Hawken, Paul 32
Helyabinsk meteor event 62
Hobbes, Thomas 65; political theory 18
Holthaus, Eric 32
homo faber, myth of 6–11

industrialization 29, 65
industrial metabolism 60
inorganic energy slaves 42
interplanetary habitation 61
Into the Extreme (Olson) 61

Jones, Van 32, 45

Kelly, Kevin 32
Kirsch, Robert E. 1, 3, 4
Klein, Naomi 29, 32, 62

Land and Sea (Schmitt) 64, 65
Lazlo, Chris 32
leisure class, Veblen's notion of 6, 11
Leviathan 65
Luke, Timothy W. 1, 2, 4, 9, 62

machine rationality 10
machinery 9, 10
Marcuse, Herbert 60, 75
market societies 4, 13
Marshall Plan 14, 17, 20
Marx, Karl 58; theory of alienation 15
material surplus 12–14
McDonough, William 32
McKibben, Bill 29, 33
Meadows, Donella 29
megamachine 5–11, 16, 19
metabolic rift 54–73
Modest_Witness@ Second_Millenium.FemaleMan_Meets_Onco_Mouse (Haraway) 60
Moon Agreement 67
Moon Treaty 56, 67
Muir, John 56
Mumford, Lewis 2; megamachine 5–11, 16; sovereign 18; theory of technics 5, 7–10
Musk, Elon 2, 3, 70

National Aeronautics and Space Administration (NASA) Center 54
natural capitalism 27
Near Earth Asteroids (NEAs) 54, 55, 59, 69, 71
Near Earth Objects (NEOs) 2, 4, 54, 55, 60, 62, 65, 71
neo-extractivism 34
NEO insurance 54
neoliberal capitalism 59
neoliberalism 57, 60, 68
neoliberalization, space 63
neoliberal space policy 67–71
neoliberal status quo 71
"new environmentalism" 28
NewSpace 68
New York Times 58
NIMBYism 44
nonproductive consumption 17
Nordhaus, Ted 25, 27, 29, 32, 46, 47
normal politics 3–4
nuclear energy 29, 31, 42, 43
Nuclear Energy Leadership Act (NELA) 28
nuclear technoliberalism 48

O'Connor, James 55, 56
"old environmentalism" 28, 29, 32, 40, 41
oligarchic corruption 70
Olson, Valerie 61
Ophuls, William 29, 42
The Order of Things (Foucault) 60
Orrman-Rossiter, Kevin 66
outer space mining 54–73; ecological catastrophe 55–60; ecological rift 55–60; limits of planetary 60–7; neoliberal space policy 67–71
Outer Space Treaty (OST) 56, 66, 67
overproduction 14

Palin, Sarah 46
Paris Agreement 77
Parson, Sean 1–4
planetary materialism 65
planetary preservationism 38
planetary spatial revolution 65
planetary urbanization 47
planet craft 35
policy initiatives 76–7
"post-environmentalism" 28, 39, 48
pragmatic approaches 27
private sector investment 67
productive consumption 3, 15–17
profitable reinvestment 3, 13
Project for a New American Century (PNAC) 32
public-private partnerships 55, 66

Ray, Emily 1–4
Rickover, Hyman G. 43
rising tide argument 66

Sagoff, Mark 39–40
Saletta, Sterling Morgan 66
Schmitt, Carl 63–5
Schneider, Steve 29
Shellenberger, Michael 25, 27, 29, 32, 36, 38, 46, 47
Silent Spring (Carson) 28
Silicon Valley 5, 58, 70; earthly gods of 2
Smith, Adam 68
social organization 2, 6, 9–11
social realm 60
social reproduction 6, 14

solar energy 43
solar excess 12–13, 18
Soleri, Paulo 38
sovereignty 18; sites of 3
Space Act, 2015 65–8
Space Colonies (Brand) 63
space colonization 57
Space Cowboy Award 63
space entrepreneurs 56
space travel 55
SpaceX 66
squanderings 13–15, 19
Stalinism 14, 20
star power 54–73
stewardship 2, 3
stored energy: concept of 1
superpower 64
surplus: consumption 12–18;
deployment 14; material 12–14
sustainable development 30

Target Earth 54
technics, Mumford's theory 5, 7–10
technological rationalization 40
technology 7; artifacts of 8
terrestrial ecological system 55
terrestrial economic system 55
terrestrial extraction 6
terrestrial limit 3

terrestrial natural resources 57
terrestrial resource extraction 59
thermonuclear exchange 20
tools 8–10
Trump, Donald 46
Tumlinson, Rick 63
Turner, Frederick Jackson 62

ubiquitous computing 39
Undoing the Demos (Brown) 68
uranium mines 34

Valentine, David 68–69
Veblen, Thorstein: leisure class 6, 11
Vogel, Steven 59

Watt, James 42
wealth 12–14
The Whole Earth Catalog 30, 31
Whole Earth Discipline (Brand)
  29, 30
Wilson, E. O. 37–8, 48
Wolin, Sheldon S. 64

York, Richard 55, 58
Yusoff, Kathyrn 60

Zahniser, Howard 56
zero-carbon technology 35